It's Time
A Practical Guide for Environmental and Social Transformation

By J.F. Hagen

~~~~~~~~~~

Edited by Myra Mueller

Library of Congress Control Number: 2016918943

John F. Hagen, Monroe, Wisconsin

Published by:

J.F. Hagen Publishing

Monroe, Wisconsin 53566

ISBN – 13-978-0-9864003-6-0

Cover photographs courtesy of Ryuzan.

Printed by CreateSpace

~~~~~~~~~~~

<u>This book is dedicated to</u>

Those who speak truth to power.

~~~~~~~~~~~

Table of Contents

Preface

I was born in the late nineteen forties. During my life, many of the most significant challenges that currently face us have developed. Having first hand recollections of events also provides a perspective on how, as a society, we have dealt with change. In American society the central cultural characteristic is greed. It is the well spring of the economic, legal and political system. The definitive characteristic of greed is that, not only do those who strongly embody this characteristic want to acquire, but also retain that which has been acquired and to continue acquiring more indefinitely. That is, they, and the institutions that are their instruments can never be satisfied. Therefore, our cultural and social systems have been and currently are extremely resistant to any form of change, particularly anything relevant to our economic system, because deviation away from these traits would inaugurate a profound reorganization of the current social structure. However, it is obvious that the way conditions are developing within our environment, it will not be possible for this type of culture to continue. The purpose of writing this is to examine the most significant cultural, social, and environmental problems facing us and to offer a few ideas and strategies that may be useful in developing a more flexible social system that has the capacity to adapt to emerging conditions. The characteristics of the new environment appear to be one of greater variation, rapid change, and depleted resources. Thus, a new social system should have the capacity to provide a reliable and adequate means of livelihood, be supportive of its members, and reduce or eliminate the destructive characteristics that currently prevail.

The way and pace that these shifts should take place are extremely important. Basically, these shifts follow three pathways. The most gentle which is being advocated, and will be presented in greater detail later in the text, could be characterized as having an evolutionary nature. It should have several characteristics: avoid being rigid, be able to coexist with the current system while its maladaptive nonviable traits are abandoned and provide a system that can be sustained without degrading or eliminating its resource base. The second possible scenario would be to wait to be overtaken by events. The probable response under this scenario, when viewed from a historical perspective, would be the formation of an authoritarian regime that would utilize draconian means in an attempt to maintain remnants of the current system. The third is similar to the second, with the difference that the current system would be rigidly maintained until a catastrophic collapse occurs. Scenarios two and three are only temporary solutions to problems that cannot be reversed and will continue to progressively decay, ultimately resulting in a more challenging survival situation. The last two types of scenarios are generally established by opportunists. The hallmarks of the methods they employ are to play upon peoples' fears and offer simplistic solutions. The people who question them are often denounced in various ways, such as character assassination, being a danger to society, i.e., pointing out deficiencies and errors in their program(s), wasting precious time, and proposing alternatives.

1. The rationales that underlies the main thrust of this book are as follows: planet earth is a large sphere, it is not infinite; therefore, the amount of resources available are definitely finite.

2. That the level of risk produced by our activities in the environment that supports us should be proportional to the potential consequences. That is, the greater the possible

negative consequences from an activity, the less it should be engaged in.

3. Greater weight will be given to relevant information that is a matter of historical record, and /or produced by direct measurement of some environmental parameter. Unverifiable suppositions, unsupported assertions, and half truths will not be used as source material.

This book will contain five sections. The first section will examine the various undesirable traits that currently prevail and how they operate in our society. The second section will examine a number of past societies that have collapsed, and some that survived, and identify how a society's response to factors that stressed them contributed to their present condition. The third section will present information on the status of the physical environment, what can be done, and what is being done. The fourth section provides a realistic appraisal of the commonly touted solutions to our environmental problems. In order to do this type of evaluation it was necessary to use quantitative comparisons, so some simple arithmetic is present to show how they were produced. The fifth section will present some thoughts on corrective measures that can be taken to remediate or eliminate undesirable characteristics in our culture and their detriment to the physical environment.

I would also like to thank all the listed and unlisted authors, researchers, internet sources, and the many other people who provided the basis of this book. I particularly would like to thank Eliza Freeman and Myra Mueller who edited this book and Alfred "Roc" Ordman who reviewed it.

Thank You All!!!!

Chapter 1
The Creed of Greed

Better to reign in hell than serve in heaven.

Paradise Lost by John Milton

Greed has been long and widely recognized as one of the most undesirable traits. Many of the religions throughout the world know that it is unwise to foster or tolerate the expression of greed, and have tenets whose purpose is to provide guidance to attenuate its manifestation. Attenuation of this trait is sought because it invariably exerts a corrupting and destructive action upon the individual and the wider society.

To start, the definition of greed will be presented, and then a survey of a spectrum of religious thought will be examined to gain a detailed perspective of its nature. According to Webster's New World Dictionary, greedy individuals are defined as possessing these traits: wanting excessively to have or acquire; desiring more than one needs or deserves; avaricious, covetous. Wanting to eat and drink too much; gluttonous, voracious. Greedy implies an insatiable desire to posses or acquire something to an amount inordinately beyond what one needs or deserves, and is the broadest term compared here: avaricious stresses greed for money or riches and often connotes miserliness: grasping suggests an unscrupulous eagerness for gain that manifests itself in seizing upon every opportunity to get what one desires: acquisitiveness stresses the exertion of effort in acquiring or accumulating wealth or material possessions to an excessive amount: covetedness implies greed for something that another person rightfully posses. I would also like to add to the dictionary's list envy, which overlaps with covetedness but also incorporates feelings of ill will, discontent, jealousy, and resentment, with an exceedingly great desire to have or acquire.

The number of warnings in religions about the expression of greed, and its more specific derivative attributes, are numerous; therefore, only a representative sample that is sufficient to exhibit the basic ideas will be presented.

Christianity

In the Christian religion a general admonition against greed can be found in 1 Timothy 6:10: For the love of money is a root of all kinds of evils; and in 1 Timothy 6:9: But those who desire to be rich fall into temptation, into a snare, into many senseless and harmful desires that plunge people into ruin and destruction. This same theme is further elaborated in Proverbs 28:25: A greedy man stirs up strife. In Jeremiah 5:26-29 a more specific enumeration of the evocation of undesirable traits arising from greed is stated: For wicked men are found among my people, they lurk like fowlers lying in wait. They set a trap, they catch men. Like a cage full of birds, their houses are full of deceit, therefore they have become great and rich, they have grown fat and sleek. They know no bounds in deeds of evil, they judge not with justice the cause of the fatherless, to make it prosper, and they do not defend the rights of the needy. Ecclesiastes 5:10: He who loves money will not be satisfied with money, nor he who loves wealth with his income, this is also vanity. John 12:6: He said this, not because he cared about the poor, but because he was a thief, and having charge of the moneybag he used to help himself to what was put into it. Luke 12:15: "Take care, and be on your guard against all covetousness, for ones life does not consist in the abundance of his possessions." All these admonishments are present in the Bible because its obvious that it isn't possible to practice the Christian ideal of brotherly love, and at the same time worship the golden calf (be a greedy person) with its attendant undesirable activities.

Native American Religions

A variety of religions exist among the Native Americans. A few generalizations can be made about how they differ from the predominant religion in America.. Many of them consider visions to be important for guidance in one's life journey. Because of the importance attached to experiencing a vision, they have various means for inducing them, which are a prominent feature of some of their religious ceremonies. The interpretations of these visions are generally performed by individuals who are considered to be adept in this activity. They also have a greater inclination to present ethical or moral standards in a positive fashion. Thus, cultivation of desirable traits about possession is emphasized. They also consider human beings to be an integral part of the natural world, and are cognizant that our existence is provided by the earth. Therefore, caring for the earth is viewed in the same light, as caring for your own body. The planet is viewed as being alive, and given great reverence.

The Native American Church of North America has the largest number of adherents, and emphasizes compassion, respect, honoring others, and sharing. It utilizes ceremonies that incorporate peyote to develop personal integrity. Visions are also considered to be important, but the use of these substances are not specifically employed

11

for this purpose. It also has adopted elements of Christianity, with particular emphasis on the commandments relating to the love of God, and to love thy neighbor as thyself. During their services, they pray to God in Jesus' name and consider his life to be one that should be emulated. As can be seen from the above, their views are the antithesis of greed. Because the tenets of Christianity are common knowledge, we will move on and consider some of the different aspects of the lesser known Native American religions. In general they advocate being generous to other people, which is the opposite of greed, and to take good care of planet earth.

The Sioux priest Black Elk[i] described the larger American society as follows: "I could see that the Wasichus (whites) did not care for each other the way our people did before the nations hoop was broken. They would take everything from each other if they could, and so there were some who had more of everything than they could use, while crowds of people had nothing at all and were maybe starving. They had forgotten that the earth was their mother. This could not be better than the old ways of my people."

In the Soulteaux Ojibwa, great stress is placed on sharing. According to Hallowell[ii]"Hoarding, or any manifestation of greed is discountenanced (P. 172)." Children are carefully and thoroughly taught that contributing and supporting the wider community is desirable. Hallowell also found that the world view of the Ojibwa recognizes that a person's actions affect the wider environment, in fact they consider themselves to be "the loci of causality in the dynamics of their universe (P. 168)."

To further illustrate the points just made, and to provide greater insight into Native American thinking and some topics that will be touched on later in this book, the following excerpts from The Essence of Hopi Prophesy[iii] are quoted here.

"The Balance of Life - As caretakers of life we affect the balance of nature to such a degree that our own actions determine whether the great cycles of nature bring prosperity or disaster. Our present world is the unfoldment of a pattern we set in motion." Later in this part the cause of our unwise activities is attributed to psychological imbalance, which is described as being a "continual struggle between our left and right sides, the left being wise and clumsy, the right being clever and powerful but unwise, forgetful of our original purpose."[1]

The Cycle of Worlds - As life resources diminished in accord with the cycles of nature, we would try to better situation through our own inventions, believing that any mistakes could be corrected through further inventions. In our cleverness, most of us would lose sight of our original purpose, become involved in a world of our own design, and ultimately oppose the order of the universe itself, becoming the mindless enemy of the few who would hold the key to survival.

---

[1]   *In modern neural psychology the left hemisphere controls the body the right side of the body which is characterized as being rational, logical, discriminating, prefers established certain information, controls feelings, and ranked authority structures. That is it has the characteristics of being clever and powerful as described. The left side of the body controlled by the right hemisphere of the brain works with hunches and is intuitive, sees connectedness has a global perspective, is fluid and spontaneous, free with feelings and prefers egalitarian social style. From the above descriptions American society could be characterized as having a predominantly left hemisphere mode of operation. The prophesy suggests that a greater manifestation of left body right brain hemisphere is needed address many of the problems that are under discussion.*

Skipping ahead through the central portion of the text to it's end: **"The Fate of Mankind"**, the text provides a description of the ideal characteristics of a society embodying a sustainable mode of existence, which is the desired outcome. The society would "be able to use our inventive capacity wisely, to encourage rather than threaten life, and benefit everyone rather than giving advantage to a few at the expense of others. Concern for all living things will far surpass personal concerns, bringing greater happiness than could be formerly realized. Then all living things shall enjoy lasting harmony."

To further probe Native Americans' views about how very intimately human beings are embedded in the life processes of planet earth, this section ends with an excerpt from a letter composed by Chief Seattle of the Squamish Indians in the 19th century about a "proposed" purchase of portions of their tribes land by the U.S. Government. " What befalls the earth befalls all the sons of the earth. This we know: the earth does not belong to man, man belongs to the earth. All things are connected like the blood that unites us all. Man did not weave the web of life. He is merely a strand in it. Whatever he does to the web he does to himself. He concludes by stating; "We are brothers after all. (see appendix for the complete text)."

Buddhism

The Buddha[2] identified three psychological characteristics that are the root causes of suffering; they are, greed, anger,

---

[2] *Buddha is a title for a person who possesses a completely perfected psychological state. The Buddha that we see statues of and usually refer to is Shakyamuni Buddha the seventh one. The practices that Buddhists engage in is to eliminate afflictive traits that cause suffering.*

and ignorance. According to Okumura Roshi[3], ignorance (self-delusion) is the root cause of the other two.[iv] When Buddhists speak of ignorance, they are not referring to mundane ignorance, such as facts that are found in a history book. They are referring to "interdependent co-origination." What this phrase means is that everything exists as a result of a nexus of causes and effects created by everything else; for example, if we take our own existence, our existence is maintained because of the food we consume, the air we breath, and the water we drink. The water comes from the oceans in the form of clouds that produce rain which we drink directly from ground sources, such as wells, lakes, rivers, and water incorporated in food. The food ultimately comes from plants that use the sun, earth nutrients, water, and air to grow. Thus, we are directly or indirectly a result of all these things. We also produce waste products that the plants require to grow, such as carbon dioxide, water vapor in our breath, and earth nutrients produced by soil organisms that consume our bodily wastes that are combined with minerals to produce soil. When we die the molecules we are composed of also return to the environment, and are reused, thereby perpetuating the cycle. Indra's Net is a beautiful way of

---

3   *There are two main streams of two main streams of Buddhism Hineyana and Mahayana. Hineyana is currently practiced by the Theravada (Teaching of the Elders) and is the more conservative branch of Buddhism, the monks are referred to as bhikku (monk the original meaning was beggar). The Mahayana arose in the 1st century CE the two predominant sects are the Tibetan branch, the monks are referred to as Lama, the other major branch is Zen, senior monks have the title of Roshi (venerable teacher). All these sects have the same goal (awakening to the true nature of existence) but emphasizing differing* approaches to attain said goal. In the discussions presented here most of the material will be from the Zen perspective.

expressing the concept of interdependent co-origination. The knots of Indra's Net are comprised of jewels, each jewel reflects every other jewel, and the net is the size of the universe; thus, everything can be said to be connected and is a product of everything else. Buddhism has a sophisticated and well developed body of knowledge about the nature of greed. In The Way to End Suffering, Bhikku Bodhi discusses the psychological mechanism of greed;[v] as arising from a cycle derived from desire that is comprised of wanting and gratification, which are not satisfied, thereby, producing pain or irritation. "To end this pain we struggle to fulfill the desire. If our effort fails, we experience frustration, disappointment, sometimes despair. But even the pleasure of success is not unqualified. We worry that we might lose the ground we have gained. We feel driven to secure our position, to safeguard our territory, to gain more, to rise higher, to establish tighter controls.....but the objects of desire are impermanent." So we see that in essence, this is a description of how grasping, covetedness, miserliness, and acquisitiveness arises. Shakyamuni Buddha found a method to interrupt the process that was just described, by employing meditation [4] (this is the seventh element in the Buddhist Eight Fold Path). Meditation develops the capability to be aware of psychological process's as they are taking place. This awareness creates enough psychological space to provide the individual with the option to choose a course of action, instead of being swept along like a loose leaf in a wind.

---

[4]   *An element of the eight fold path which takes one from suffering, it is comprised of: Right understanding, Right attitude, Right speech, Right action, Right livelihood, Right concentration, and Right awareness. It is the fourth of The Four Noble Truths, i.e.,Their is a means to end suffering.*

The remaining seven elements of The Eight Fold Path either serve to avoid or act as antidotes to the defilements of the mind. For example, generosity is considered to be a desirable trait. If practiced diligently, it acts as an antidote to greed. Actualizing desirable traits acts as an antidote, because our consciousness is considered to be comprised of 2 strata, the conscious part that we use for every day activities and a store conscious where a complete collection of traits reside, generosity and greed being among them. The strength of these traits depend upon how often they have been aroused. If they are aroused frequently, they become stronger, and if not aroused they diminish as forces within our personality. Thus, the Eight Fold Path is a complete method for psychological improvement. It provides a comprehensive description of desirable psychological attributes, and also a means to enhance these traits as well as diminish the afflictive elements of our personality. By implementing this, it is possible to reduce and ultimately eliminate the undesirable actions that produce suffering in our and others' lives. To put it more simply, it provides a map and a means that can be utilized to reduce suffering.

The development of compassion and wisdom are primary goals in Buddhism, along with generosity, actualizing harmony in the community, and to manifest truth.[5] Like the native American religions they tend to emphasize a positive approach by stressing the actualization of desirable traits.

---

[5]   *The last three are part of the sixteen precepts taken as vows for ordination. Since Buddhism is not based upon revelation they are not considered to be commandments, but nonetheless persons who take these vows do their best to carry them out.*

Considering the description of the mechanism of greed by Bhikku Bodhi, greed embodies many of the same characteristics as addiction, an overwhelming craving for something. The craved object(s) are superfluous to or exceed any real need, and therefore, the needs cannot be satisfied. Because of its unbounded nature it has a host of the most pernicious effects. The greedy individual continuously exerts a destructive corrupting influence on themselves, others and society. At the same time they needlessly consume and degrade the physical environment through injudicious actions. As has been illustrated in the discussions and examples presented above, all the religions that have been discussed condemn it! To recapitulate a few examples: Christianity in 1 Timothy 6:10 clearly condemns it while advocating its opposite of brotherly love. It's demonstrated in the Native American religions by their persistent efforts to teach children to avoid hoarding and being greedy, and to share with others. Buddhism considers greed to be one of the three poisons of the mind that produces suffering. Buddhism and the Native American religions also recognize that the extractive behaviors that generally accompany the manifestation of this trait is also highly destructive to the wider environment of which we are a part, as illustrated by Indras' Net and Chief Seattle's web of life.

Big Business

To start, the underlying social and psychological factors that give rise to the laissez-faire capitalist system will be considered. Ownership, prestige, and status are the principal socially derived drivers that provide suitable conditions for the existence and operation of this system. Ownership has two aspects: it provides a means of eliminating conflict by defining usage rights, and establishes possession, an exclusive form of control. By its exclusive nature it provides the means for excessive acquisition, enabling the expression of greed. In addition to excessive acquisition it also provides a system for the denial of resources to others. In order to make the extractive forms of acquisition palatable to the American polity, an ideology of exalted individualism and excessive competitiveness are fostered. This ideology provides the basis for acceptance of the unequal distribution of resources, by allowing personal success to those who conform. For success in overcoming rivals is the preeminent path for the acquisition of prestige and higher status, which confers many advantages.[6] These advantages are the acquisition of wealth, prominence, desirability, and social advantages. In order to enhance the individuals' chances to achieve success a number of supporting characteristics are fostered; craftiness, guile, aggressiveness, callousness, self-importance, and ruthlessness. Moreover, the scale, structure and social organization in America produces individual anonymity and social insulation which are essential enablers of the current extractive economic system. Those who do not embody a pronounced array of these traits are winnowed out and have few chances of attaining of elite positions.

---

[6]   *Prestige is generally earned and can be acquired by other means then wealth. For example, Mahatma Gandhi was not a wealthy person and the acquisition of wealth was not a driving force in his activities but he had immense prestige. Status refers to a formal social position. A person can have high status but low prestige such as a Governor of a state that gets caught engaging in corrupt activities.*

Taking a closer look at the globalized capitalist economic system it is apparent that a close correspondence exists between the way it operates and the way the mind functions for most people. The typical mode of operation of the mind is to continuously occupy itself with dwelling on the past, having fantasies, and making plans for the future, based upon the fantasies and recollections of the past. The nature of the recollections are conditioned by past experiences, and are a partial often distorted representation of what actually occurred at the time of formation. The distortions and selectivity of these memories and associated emotions are a result of the interaction of cultural training, family traditions, and the way a person's fundamental traits are constellated into character. As is often the case, actual events do not correspond with the plans that were formulated, since they were based upon flawed assumptions. The ego uses a number of strategies to cope with this problem discrepancies are ignored if possible. If not, then a gloss is given to the unexpected contrary events. If the first two strategies fail, people will occasionally adjust their basic assumptions. If we compare this style of thinking to the basis of the capitalist system of business the same underlying pattern is present. It operates by starting with fantasies based on conditioned recollections of the past, which are then embodied in some type of plan for the future. Generally the plan presents a distorted view that is limited to a small, selectively chosen set of assumptions that are used to make linear projections. The plan provides the impetus to attract resources in order to provide the means of fulfilling the fantasy of the investors (owners). If the plan is successful in acquiring wealth, the fantasy is temporarily fulfilled. However, since these fantasies are based upon persistent underlying psychological drives, and are constantly being reinforced by society to maintain or enhance their presence in the personality, there is no mechanism to produce a holistic viewpoint congruent to prevailing conditions. Thus, a psychological climate that produces excessive acquisition and possession of material objects is established. Therefore, greed is essential to support laissez faire capitalism, and to provide psychological comfort for the participatory elites.

In American society there are several kinds of social controls to regulate deviant behavior an informal system that acts through effecting changes in prestige and/or status, various types of shunning, and a legal system that utilizes various methods of compulsion. 6 The legal system is fairly rigid. Being based upon precedents, it has only a circumscribed ability for adaptation. For example, the Ford Motor Company had a vice president and stock shareholder named Mr. Couzens, who apparently started to have attacks of conscience about the treatment of the company workers. The plant workers were receiving low wages of $2.34 per day for 9 hours of work. Also at that time the number of cars that were sold in a particular month varied, being particularly low for December. The company's policy was to lay the workers off when demand was low, which of course produced much hardship, particularly during the winter holidays. In 1913 he proposed that the workers be given a raise to $5.00 per day for 8 hours of work, more than doubling their wages. He also proposed that the company reduce the number of layoffs, particularly during the December holidays. This was not well received by Henry Ford, who vigorously tried to dissuade Mr. Couzens from implementing improved worker benefits. Mr. Couzens was a man of integrity though, and could not be dislodged from a position he felt in his heart was right. Mr. Couzens did fully and promptly implement the proposed changes on 5 January 1914, and shortly thereafter he was forced to resign from the company. His biographical information indicates that he never regretted his actions.

The increases in worker benefits raised quite a howl among the other wealthy industrialists. They ran a robust public relations (PR) campaign comprised of numerous articles in the press decrying the new policy, and demanding that it be rescinded. This same group of industrialists also would not engage in social contact with Mr. Ford. The new Ford policy was quite popular with ordinary people though, and gave a considerable boost to Ford sales. Thereby, demonstrating an attempt to change his undesirable apparent non-greedy behavior through the use of informal means of control. At a later date Mr. Ford, who owned 68% of the company, wanted to upgrade the company by using earnings produced from sales, thereby diminishing the amount of dividends. The Dodge brothers, who were also share holders in the company as well as a major vendor, filed a law suit against Ford for not paying the maximum amount of dividends. Ford lost this suit in 1919 and the legal decision read as follows: "But it is not within the powers of a corporation to shape and conduct a company's affairs for the merely incidental benefit of shareholders and for the primary purpose of benefiting others." These incidents clearly demonstrate that incorporated businesses sole purpose is to maximize profits and that greed occupies a central cultural position.

As just illustrated, the sole purpose of a corporate business is to produce a maximum amount of returns to the owners. To provide ever increasing profits, market development is a major preoccupation. The development of a market typically follows this pattern: a new market is created which is partially developed to some extent by the originator, and further expanded by other opportunistic businesses. Eventually saturation occurs, achieving its sales potential. Once the market potential has been reached the only options open for increasing profits, are through business expansion by increasing market share, and/or reducing costs. The means of expansion of market share that are available in a mature market are to buy up the competition (a merger), or force competitors out of business. Ultimately the objective of a capitalistic enterprise is to achieve a monopoly, thereby eliminating a free market for the product or service.

Let's consider some of the means of expansion in greater detail. The product or service can be improved in some way or offered at a lower price. Promotional methods to produce an emotional basis for purchase of the product or service is a commonly used technique to enable higher pricing, and thereby produce greater profit. The most common method of producing emotional appeal is through incorporating a magical feature. The magical feature induces a perception that the product can act as means to relieve feelings of, anxiety, envy, or inferiority. Anthropologists have identified two ways by which magic operates. Magic works through the perception that some property can be transferred by association with an object, and/or having shared similarities to something. Advertising psychologists are well aware of these mechanisms. For example: the magic of similarities is commonly used to induce buyers to purchase high-priced tennis shoes by incorporating similar features to ones worn by a celebrity athlete. The magic of association is particularly important in collectable items. For example, if you were a guitar player/collector, Elvis Presley's guitar would command a much higher price than one of the same model and condition that had been owned by Joe Schlunk. These same mechanisms are used to create "special characteristics" for branded, designer, or other feature(s) that will produce an image, usually to enhance the possessor's prestige, status, and/or self image or persona.

Superior performance in the case of product innovation usually results in only a temporary increase in market share, because the same novel feature or a similar one can be copied and incorporated by other producers of the same type of product. A few examples are automatic transmissions in vehicles, high definition TV, cellular telephones, etc. All these innovations were quickly and widely adopted throughout their respective industries. The other type of innovation occurs within the process to produce a product. This type of innovation is not obvious and may be kept secret for some time. For example, the power loom for making cloth was invented in England. To preserve the commercial advantages of this technology, English law prohibited the transfer this technology to other nations. Of course what eventually happened was that a person with a knowledge of how to make the power loom immigrated to America. The forbidden technology was quickly recreated, and rapidly diffused into the American domestic textile industry.

Psychological Methods of Market Expansion

A widely used means to produce product demand is to create or amplify feelings of anxiety. These feelings are used to induce a desire to purchase something to provide relief, such as various types of insurance, protective devices, and wars. Wars are very lucrative for many industries; they require and rapidly use up huge amounts of equipment and resources (we will return to this topic in greater depth in the next section). An illustration of how anxiety is used for these purposes occurred during the Nuremberg trials of Nazi war criminals. Herman Goring was asked how the Nazi regime convinced the German people to accept its program of conquest. He replied that it is very easy to do; all that needs to be done is to create fear within the populace. Anyone who questioned the militant policies and propaganda would be denounced as weakening the war effort and being a danger to the state and the nation. The question this raises is, why was it so easy to convince the German people to accept what in retrospect was utter folly? Neural scientists have discovered that when one feels threatened, the apparent threat is processed by a part of the brain called the amygdala, where the fundamental survival mechanisms reside. It has the capacity to override the higher analytical functions of the mind. It dominates because the amygdala has large numbers of one-way neural connections to the neocortex, but few that go from the cortex to the amygdala. Also the nerve that goes from the eye to the amygdala is shorter than the one that goes to the cortex, thus the amygdala also receives information and can act on it before it arrives in the other area of the brain. Another physiological response to fear is for the blood supply to be redirected to other parts of the body, away from the parts of the brain where critical thinking occurs. The result is that a person when exposed to this type of stimulus will follow the content of a persuasive message relevant to the apparent danger in an attempt to neutralize it. Not only are politicians aware of this behavioral mechanism, but also the advertising industry. For example, SARS, a potentially fatal flu virus, appeared in 2003, and later in 2009 an outbreak of the virulent H1N1 influenza virus occurred. Industry quickly launched advertising campaigns designed to capitalize on fear of contracting these diseases to sell products such as antibacterial hand gels, soaps, disinfecting wipers, and protective medical kits. Since these diseases are transmitted by airborne viral particles, these products have no effect on preventing these diseases. Of course the ads were carefully designed not to directly say they would prevent these diseases, since that would be a blatant lie, but composed to suggest or imply that they would, by using the magic of similarities to other pathogens. Not only does the advertising industry capitalize on existing events, it also creates imaginary or exaggerated fears, as do politicians. The use of fear to shape public opinion about a number of industries in the United States has led to extreme distortions of most peoples' perceptions. Let's compare some of what are considered to be very scary industries and relatively benign ones. If we compare the safety records of commercial nuclear power generation and "clean coal," these are the facts: the amount of radiation produced by a coal-fired electrical generation plant is 100 times more than a nuclear plant (because radioactive substances are present in coal). In fact if coal-fired electrical plants were held to the same standards for radiation as nuclear plants, they would all have to be shut down! If we compare the number of deaths that have occurred in each type of plant we find that no deaths have ever occurred from commercial nuclear power generation in the United States, whereas coal currently produces around 31,400 deaths per year.

The coal industry PR is that it's inexpensive and that using it is an economic necessity. Is this true though? A recent study conducted by the Harvard School of Public Health found that if the externalized costs engendered by the use of coal were included in the price of electricity, it would increase its cost by $.18 per kilowatt hour, making it one of the most expensive sources of energy. That is between $150 billion and $500 billion are paid each year by the public through higher taxes, property damage, greater insurance rates, health care costs, etc. (Note: In the next section we will consider this in greater detail).

Short Product Life - A prime objective in the type of economic system that currently prevails is to produce products that require frequent replacement and induce continuing usage. To accomplish this, products are designed for one-time use (disposable products originated my interest in this topic and will be discussed below), planned obsolescence and upon addiction, to maintain demand after attracting and inducing a person to start using it. Carmex lip balm illustrates this point. Their product induces a need for continuing usage by incorporating a combination of ingredients that produces an effect requiring further use of the product. Lyndstrom quotes Dr. Pray's description of its activity as follows: the ingredient phenol which acts as "a deadening agent that literally anesthetizes our lips, at which point the salicylic acid begins eating away at our living tissue, namely our lips."[vi] The incorporation of these ingredients apparently is designed to produce an ongoing need to continue using the product (one of my friends calls this a self-licking ice cream cone). As another example, the tobacco industry frequently promoted their addictive products using celebrities, cool images in teen hang-outs, and cartoon images for younger people to attract new users. New users of tobacco quickly become addicted to this hazardous product, thereby, becoming the ideal continuing customer. The computer and software industries very obviously are based upon planned obsolescence. A typical laptop computer is only expected to have a two year life expectancy. Much of the software also cannot be used on an "older" machine, thereby, requiring continuing purchases. Let's consider computer viruses. Anyone who uses the Internet has problems with these, which always require one to obtain a state -of-the-art computer virus removal software. It's interesting if one considers these viruses. They are obviously produced by individuals who have a comprehensive and current knowledge of how computer viruses are made and introduced into computers through transfer on the Internet. Who would have the capability and motivation to produce and introduce computer viruses? If one considers a financial motive (greed), an obvious place to look is the very businesses and individuals who produce virus removal software. They are certainly dependent upon the appearance of new computer viruses, have the technical capability and means of introduction. Well of course this is just unproven speculation, and one can't say for sure. After all, a possibility does exist that all the owners and people who man these enterprises are paragons of moral rectitude, who would never engage in these practices; perish the thought.

Monopolistic Trends

If one considers where the process of constant growth of a business ends, it ultimately results in the formation of a monopoly. Since the United States has laws designed to prevent the formation of monopolies how can a multinational corporation achieve this outcome?[7] Simply by producing their product or service outside the United States.

This business practice has become common in the last several decades. For example, starting in the late 1980's a lot of debate in the U.S. took place about granting China most favored nation trading status. It was delayed for some time because the U.S. government had a policy of not granting this status to nations with a record of human rights abuses. China having an appalling record, i.e., the Tienanmen square massacre and the ongoing persecution of the Tibetans since they were conquered in 1959, precluded them from this status. In the early 1990's an advertising campaign was conducted to change public opinion in order

---

[7]   *It is interesting to note that large agribusinesses are apparently immune to the antitrust laws.*

to reverse this policy. It was based upon arousing greed, it was argued that China was this vast market for the export of American products. The PR message was that everyone would financially benefit through profits derived from sales and the creation of many high paying domestic industrial jobs. At the time I was mystified how this could come about since the average Chinese worker made less than $20.00 per week and an engineer around $75.00 per week. The question was, where would they get the money to buy $40,000.00 SUV's, $500.00 appliances or $100.00 tennis shoes when the vast majority were barely surviving? Some time later after a number of the multinational corporations transferred their manufacturing to China, it became apparent the main reason was to provide a means for these large multinational businesses to take advantage of weak or non existent labor laws, environmental regulations, and very low labor costs to undersell and force their domestic competitors out of business thereby gaining market share. Of course the transfer of manufacturing offshore caused huge numbers of high paying American industrial jobs that already existed to be lost! Again, to blunt public concerns a PR campaign was launched. The PR extolled the virtues of working in service positions instead of in what was disparagingly referred to as being dirty out of date industrial positions in rust bucket industries and other negative tendentious monickers. One can imagine how happy the displaced workers were when they moved into the new "nice clean service positions" which on average pay 20% less than the industrial jobs they replaced. Most of the service positions also lacked the valuable benefit packages that usually accompanied industrial jobs.

As just mentioned, some of the reasons for relocating their businesses offshore was to evade paying for workers benefits and environmental regulations. This is referred to as externalizing costs. Externalizing costs is a euphemism used to describe the business practice of avidly seeking ways to foist off as many of their direct and indirect costs onto the community forcing others to subsidize their enterprises.

Let's consider some of the costs of the use of disposable products. When we use these products, throwing away a container, or other disposable item such as a pen seems to be trivial. However, if one considers the amount used collectively, it is not. For example, after graduating from high school I obtained a job with a company that produced house-ware products. The production of disposable aluminum containers for food was a major component of their business. The job that I had was part of a two man team that operated an aluminum shear to size the foil used to produce the containers. The company had two larger machines and one small machine used for thin foil such as you use to wrap things. A large machine on most days would be able to shear approximately 20 tons of foil in eight hours and the small machine several tons, thus, in one shift the combined output was around 40 tons. These machines were operated for three shifts a day producing around 100 tons of sized foil that would be promptly converted into disposable foil products, or about 500 to 600 tons per week.

When usage is considered on a yearly basis, around 29,000 tons where used (we often worked 6 days). This was before recycling so all of this aluminum was going into dumps. The colossal waste and other environmental problems associated with this type of disposable product was troubling to me, and ultimately produced the motivation to start investigating these types of issues.

Starting in the 1950's a great expansion of the development and use of disposable products took place. In my youth, it was a common activity for youngsters to get spending money by scouring the area where they lived in order to recover discarded reusable beverage bottles. These containers could be returned to a supermarket, liquor store, etc., to recover their deposit. The refunds on the bottles were: a 12 ounce bottle $.02, 1 quart $.05, and 1 gallon $.15. To put this in perspective at that time a cup of coffee in a restaurant was $.10, few bottles went to waste. In the 1970's the people who were environmentally aware were cognizant of how needlessly wasteful our society had become. To reduce the high level of waste it was decided to attempt to rectify the situation by regulatory means. The idea was to return to the use of improved types of reusable durable containers. What was desired was not only to use reusable containers as before, but have national standards for them that would specify their shapes and sizes. The use of standardized containers was sought because it becomes feasible to sort and clean them using automated equipment. This would have greatly reduced waste of materials, energy, and the cost of packaging born by the consumer. Studies of the proposition demonstrated that further benefits would result because a net expansion of the number of jobs would take place as a result of the need for people to operate the new system. This idea was not welcomed by the disposable container industry. Of course it also gained no traction with the politicians who received their financial campaign support from these industries. A solution was sought that was more palatable, and thus the idea of recycling was born. Recycling is a lot less effective than standardized improved durable containers for resource conservation. The container industry, material producers, and the politicians were still adamantly opposed to any change away from the disposable products. The folks in the environmental movement decided to pursue the second option (recycling). This approach reduced the amount of resistance being produced to that originating from the material producers who would feel the greatest effects.

Therefore, a nationwide petition drive was organized to place it on the election ballots as a referendum in order to weaken political resistance. In Illinois $50,000.00 was raised for the petition drive. The various industries spent $3,000,000.00 to oppose the referendum, these funds were spent on lobbying, political contributions, and PR. The initiative was defeated in Illinois and even if it had been a success, referendums in Illinois are only advisory and non-binding. Non-binding referendums have no legislative force, thereby allowing politicians to ignore successful referendums if they choose. However, the initiative was successful in some other states (you can see a listing on the bottom of beverage cans).

How have things progressed since then? The amount of packaging and use of disposable products have greatly increased. Recycling has proved to be only partially successful. For example, only 27% of the hundreds of millions of plastic water containers are actually recycled; one hundred million disposable pens are used each year and discarded. The amount of plastic that can be recycled is limited because it loses its engineering properties. A common specification for the production of plastic items is to allow a maximum of 20% regrind (reprocessed used material). Many products require the use of 100% virgin material. The fibers in paper products such as cardboard break down when being reprocessed and also lose their properties limiting the number times of reuse. Metal's properties do not breakdown, however reprocessing metals requires a lot of energy, although much less than needed to convert mined ore into new metal.

Symbiosis of Government and Wealthy Interests

In the United States our electoral process is quite expensive (the 2012 election cycle cost $6 billion), and is dependent upon private funding. These private donations in some cases can be augmented by government matching funds. The matching government funds are available to those who have been able to attract 5% or more of the vote and are released after the election. Since independents and third party candidates campaigns are generally poorly funded and have no staying power, the lions share of matching funds go to politicians sponsored by one of the established parties.

Lets consider one of the largest political campaign contributors and sources of lobbying, the armaments industry. After world war two, the United States established a policy of maintaining a large standing peace time military establishment. President Eisenhower was aware and concerned about the formation of what he referred to as the military industrial complex. He consistently sought to restrain its growth by restricting the amount of military spending. President Eisenhower gave a number of speeches on this topic, the last one shortly before he left office in 1961 which presents a fine portrayal of the problematic effects the relationship between industry and government was being produced by this policy.[vii] He noted that prior to the cold war the United States did not maintain a large standing army supported by a large permanent armaments industry that required large scale scientific and engineering support. He warned that "we must guard against the acquisition of unwarranted influence, weather sought or unsought, by the military industrial complex. The potential for the disastrous rise of misplaced power exists and will persist.......domination of the nation's scholars by Federal employment, project allocations, and the power of money is ever present and is gravely regarded." The final points that were made in this speech warned that there would be an "impulse to live only for today, plundering, for our own ease and convenience, the precious resources of tomorrow." He concluded that it was imperative to "avoid becoming a community of dreadful fear and hate".

Well here we are in 2013, how have we done? In one of Eisenhower's other speeches he pointed out the tradeoffs being made to support a large military establishment the gist of which I shall update as follows: currently a new aircraft carrier the Gerald R. Ford is being constructed at a cost of $13.5 billion dollars. If this money were spent on schools (the average cost of a school in 2013 was $20.6 million) 655 new schools could be built. A B-2 stealth airplane costs $1 billion, or 48 schools could be built.

The United States military budget for 2013 was $716 billion. When our politicians are questioned about the need for this level of expenditure they usually reply that it's needed to provide adequate equipment and training for our military forces. Is this true though? Lets consider the relative strength of our military. The United States nuclear arsenal is currently comprised of 7,700 bombs of which 1,950 are strategic bombs that are deployed, 200 nonstrategic bombs, and the rest are in storage. During the cold war the U.S. government evaluated the number of potential targets where a nuclear bomb would be useful and found that the number was around 200. These numbers suggest to me that the likelihood of being attacked by another nation seems very remote since they would be facing certain annihilation if even a tiny fraction of these bombs were used. If we compare the size of our military expenditures to the next highest nation which is China. China was spending $106.4 billion in 2013, and that the U.S. is spending 6.7 times more ($610 billion more). The

U.S. navy is larger than the next 13 largest navies combined, 11 of these navies are possessed by our allies. By 2020 the number of stealth airplanes we are building will be 20 times what China will have and also be much more technologically advanced. The addition of the Gerald R. Ford will bring the number of deployed aircraft carriers up to 10. Currently only one other nation has an aircraft carrier and it is of much smaller size. It is owned by the British which is one of our allies. When considering these facts it is obvious that the United States is not maintaining a military appropriately sized to act as a defensive deterrent but one to support an aggressive foreign policy which may be characterized as a form of economic imperialism. If we recall the reasons for the formation of the United States, the American revolution occurred as a result of economic exploitation by a colonial power. If we look at recent history it seems that the United States government is engaging in the same kind of policies that the country was originally formed to escape from! What Eisenhower was warning about, the formation of a military industrial complex was anticipated as a potential problem by the framers of the United States constitution. The Federalist Paper Number 10 (22 November 1787) articulated an expression of deep concern about the potential for the formation of factions that would come to dominate the political process with narrow interests contrary to the interests of the wider community. United States Supreme Court Justice John Paul Stevens cites Federalist #10 as saying "Parties ranked high on the list of evils that the constitution was designed to check". James Madison also wrote in Federalist #51 on the 6 February 1788, that political factions will always be present and the way to counteract excessive influence by a small number of factions was to have numerous factions. idea was that no group could become strong enough to thwart the interests of the other groups.[8]

In light of these comparisons I would argue that Eisenhower hit the nail on it's head and his predictions and the founders fears have been realized (a further discussion of this will be presented in the last section). In my estimation the United States military is extremely over built.

---

[8]   *Prior to and during the revolutionary war the use of propaganda (or public relations [PR]) campaigns was very much in its infancy being confined to paper publications and public speeches. During world war II governments used the new mass mediums of movie theaters, radio in addition to the older use of printed materials. Moreover, the public speeches during this period were using electronic amplification and could reach much larger crowds of people materials. By the time Eisenhower made his speeches Television had become an important means of disseminating information. As can be seen a progression in the pervasiveness of the mass media took place allowing ever greater means of shaping public opinion. If we consider the last twenty years or so this trend has continued to take place up to the point where people are completely saturated with these types of messages. Their is a big difference however, not only has the presence of this material become much greater through the use of individual portable electronic communication but the effectiveness is far beyond what had been previously achievable. The use of recent psychological discoveries of how the brain works coupled with the use of f MRI to see which portions of the brains of test groups are actuated to craft advertisements and propaganda has produced exceedingly effective PR instruments for shaping the targeted groups acceptance of the presented material. The framers of the constitution certainly had no idea about this when they were building in the checks and balances in the United States government. Even Eisenhower when he gave his speeches was only seeing a very feeble expression of the current capability when he gave his speech.*

This situation could be rectified if the funding that governs its size were scaled to the amounts being spent by other nations, perhaps a ceiling of 120% of the next largest. If the arithmetic is done for this level of funding, our budget would be around $128 billion freeing up $588 billion of public funds that could be used for other purposes such as education, infrastructure, research, medical care, etc. Not only would a colossal amount of funding become available for directly beneficial purposes but it would also increase our security. If the numbers of nuclear weapons are considered they pose a grave threat to life on planet earth. American, Russian, and European climatologists have all warned that if around 200 megatons of bombs were exploded the large quantity of aerosol particles and combustion products would alter world wide temperatures and cause a protracted climate change. This type of alteration in the climate would impair the growth of plants eliminating the source of food for human beings and animal life. Moreover, the aerosol particles produced by nuclear bombs are comprised of radioactive fallout which would cause widespread contamination of the earths surface and water. Barbara W. Tuchman[viii] performed an extensive historical analysis of a number of wars embracing the period from Troy to Vietnam. The results of this research enabled her to identify a number of typical types of flawed thinking that occurs by those who are in power. One of these is the propensity for politicians to use force if they feel confident that it will succeed. Another source of folly she identified is to trade the greater for the lessor. If we consider the recent wars in the Middle East it seems unlikely that they would have been undertaken if the distribution of military potential was not so asymmetric. With respect to the second source of folly, if the trillions of dollars [9] spent on these wars, loss of world standing, and the corruptive nature of these activities upon American society are compared to the benefits accrued to our nation as whole, there is no doubt that the greater was sacrificed for the lessor.

Lets consider The United Fruit Company [UFCO] as a case study of how commercial enterprises can manipulate our political system to provide public resources for their own purposes. UFCO originated from mergers of three companies, The Boston Fruit Company started by Captain Lorenzo D. Baker in 1885, International Railways of Central America founded by Minor Keith, and Samuel Zemurray's fruit company. After the merger UFCO possessed extensive land holdings throughout the Caribbean and Central America. Historically the general mode of operation of the three root companies was to make deals with a dictator in control of a country in order to gain concessions, usually for reduced tax liabilities, below par land acquisitions, and control of labor costs (in Guatemala workers were paid $0.50 per day). These concessions were obtained through various corrupt means. Thomas McCann who worked for UFCO for about 20 years summarized its Guatemalan enterprise as follows: "Guatemala was chosen as the site for the companies earliest development activities

---

[9]   *A Harvard University study estimates that the total cost of the Afghanistan and Iraq wars will cost the United States taxpayer $4 - $6 trillion dollars when the medical and disability costs are added to the $2.3 trillion in direct military expenditure. In addition to our own military deaths, the October 2006 medical journal Lancet estimated with a 95% confidence level that there has been 654,965 war related civilian deaths in Iraq. This doesn't appear to be the "surgical military operations" with "low collateral damage" that was presented by the United States government in its press releases.*

at the turn of the century because a good portion of the country contained prime banana land and because at the time we entered Central America, Guatemala's government was the weakest, most corrupt and most pliable. In short, the country offered an "ideal investment climate," and United Fruit's profits there flourished for fifty years. Then something went wrong, a man named Jacobo Arbenz became President."[ix]

Arbenz described UFCO's activities in Guatemala as "All the achievements of the Company were made at the expense of the impoverishment of the country and by acquisitive practices. To protect its authority it had recourse to every method; political intervention, economic compulsion, contractual imposition, bribery and tendentious propaganda, as suited its purposes of domination. The United Fruit Company is the principal enemy of progress of Guatemala, of its democracy and at effort directed at it's economic liberation." The American historian Cole Blasier wrote in the Hovering Giant: U.S. Responses to Revolutionary Change in Latin America, "In the past, UFCO and it's sister companies had bribed politicians, pressured governments and intimidated opponents to gain extremely favorable concessions."[x] If we consider the activities of Mr. Zemurray prior to the merger of the three companies when he became CEO of UFCO the types of business practices just described above were normal for him. In 1905 he wished to expand his business in Honduras where he planned to buy land, build a railroad to the coast and obtain concessions. The concessions he desired were comprised of a low rate of taxation that would be frozen and a exemption from import duties. At this time Honduras was bankrupt and President Magill Davila was negotiating a bailout deal with a New York bank. One of the conditions for the loan was that an agent of the bank would take control of the treasury. Mr.

Zemurray found out about this and realized that the concessions he desired would not be available if an agent from a bank was in charge of the treasury. He decided that a person who would be sympathetic to his business goals was needed and found an enemy of the incumbent president named Manuel Bonilla who he outfitted with a surplus navy ship (the Hornet) and additional armaments suitable for land warfare . Shortly thereafter a revolution took place in Honduras and Manuel Bonilla became president. President Bonilla gave Mr. Zemurray extremely favorable concessions. Later in 1915 Mr. Zemurray bought up large tracts of land in the Motagua valley along the boarder of Guatemala. In 1930 UFCO bought out Zemurray's complete holdings for $31.5 million in UFCO stock making him the companies largest stock holder. In 1933 he used his stock to obtain the position of managing director of the company.

In the early 1950's the company became aware that things were changing, the impoverished and disenfranchised populous was no longer satisfied with the status quo which created a climate where agents of change could flourish. The result was the election of the Jacobo Arbenz government that enacted legislation and started a program to attenuate domination of the country by UFCO. The program instituted was to build a road that terminated at the Atlantic coast, and an electric power plant, thereby, eliminating the monopolies possessed by American interests. At this time UFCO's transportation monopoly was comprised of the railroad (IRCA) and the only Guatemalan Atlantic port at its terminus Puerto Barrios. Ownership of Puerto Barrios also provided UFCO with the further advantage of being able to restrict the handling of sea freight to its own ships (it owned 50 ships for this purpose), or on infrequent occasions to outside shipping vendors it approved of. According to the International

Development Bank the freight rates that were being charged were the highest in the world. The situation that existed would no longer support a staged coup as was implemented in 1905. Therefor, UFCO's management decided to induce the United States Government to use its power to maintain its corporate hegemony within the country. In order to accomplish this task one of the best PR men of the time was hired (Edward Bernays) and also a number of well connected liberal lobbyists. These lobbyists had contacts in the CIA the senate, congress, and the State Department notably, John Foster Dulles, Secretary of State, his brother Allen Dulles Director of CIA, as well as an earlier director of the CIA Walter Bedell Smith, who later was Under secretary of State with Dulles.[10] After Dwight Eisenhower was elected to the presidency Mr. Zemurray realized it was now essential to expand his influence into the right wing faction of the Republican Party, again he engaged a well connected PR firm, John Clements and Associates. Mr Clements was also Vice President of the jingoistic Hearst Publications. He was also editor of The American Mercury a McCarthyite propaganda sheet as well as being related to Senator Joseph McCarthy.

The strategy that UFCO employed to garner support and protection of it's overseas interests by the United States government was to create and amplify fears about the spread of communism. The basic underlying theme they used was the Guatemalan governments program of land reform.[11] The land reform program was used by UFCO's

---

[10]   *After Walter Bedell Smith left the U.S. Government he obtained an executive position with UFCO.*

[11]   *UfCO owned 550,000 acres in Guatemala it had 15% of this land under cultivation initially the Guatemalan government wanted approximately 209,000 acres and later upped it to 386,901*

PR people to create the fantasy that it was engendered by communists supported by the Soviet Union. The public relations campaigns employed for this purpose were comprised of two elements. The first element was aimed at the general American public to create a supportive climate that was essential for governmental decision makers to operate in. The second was specifically targeted and tailored for U.S. government decision makers. The PR effort targeted on the decision makers was augmented by intensive lobbying. During the Truman administration the PR effort was comprised of a "confidential" news letter that was circulated to 250 journalists, congressmen, institutions such as the Council on Foreign Relations, and congressional offices.

The press PR effort was augmented by free junkets for newsmen of prestigious publications such as Time, Newsweek, Scripps-Howard newspapers, United Press International, San Francisco Chronicle, Miami Harold, US News and Word Report, The New York Times, Atlantic Monthly, and in some Latin publications.[xi] UFCO produced 4 or 5 of these junkets a year taking up to 10 reporters each time. They were very carefully orchestrated to present a favorable impression of the company and support it's assertions of benevolence. No expense was spared on these PR junkets in one year $500,000.00 was spent on these which was a very large sum at that time (in 1950 the average price of a house was $8,450). Dozens of officials in the U.S. Congress and Senate were induced to act in a favorable manner to UFCO, they often read

---

acres, the company was offered compensation equivalent to the land values it provided on it's tax returns. At this time UFCO was paying $150.000 per year in taxes, 1 cent per banana stem and nothing on its property.

company prepared PR on the Senate and Congressional floor and into the congressional record. After the Eisenhower administration took office, UFCO retained John Clements who had a list of 800 influential decision makers. In 1952 Clements was paid $35,000.00 to write a 235 page anonymous "study" titled "Report on Guatemala - 1952" for UFCO. This document was bound in a manner similar to that used for official documents and sent to the 800 people on his list. Excerpts from this document later appeared in speeches presented at the UN, a State Department White paper, and other official documents. Clements was also a friend and business partner of Nicaraguan President Somoza. In 1954 UFCO's efforts bore fruit, the democratic Arbenz government fell as a result of US engendered attacks[12] from Honduras, and was replaced with a dictator.

Upon examination of the examples already given we can characterize several traits that are common to a greed based system. The individuals controlling these large enterprises embody a vision that is restricted to activities primarily focused on maximum gain and ignore the wider impacts of their actions. In order to accomplish this goal every means is used: co-opting of public resources, externalizing costs, manipulation of the political/regulatory system, exploitation of its working force, and the direct or indirect use of force. One of the primary tools used to produce a climate suitable for gaining these ends is to produce PR that

---

[12]   *UFCO provided training and staging areas on it property for the "liberation army" as well as $64,000 in cash. The aftermath of their activities were decades of military dictatorships that maintained their position through a continuous rain of terror with an estimated 150,000 - 200,000 persons killed.*

is often disingenuous. In his 1928 book "Propaganda" Mr. Bernays described the goals of these PR campaigns as "The conscious and intelligent manipulation of the organized habits and opinions of the masses is an important element in a democratic society. Those who manipulate this unseen mechanism of society constitute an invisible government which is the true ruling power of our country.....it is the intelligent minorities which need to make use of propaganda continuously and systematically".

If we consider the current status of how the powerful vested interests are utilizing the techniques of PR to shape a favorable climate for gleaning profits and power, their presence has become much more pervasive and effective because of the ubiquity of electronic communication, and advances in the psychological and neuropsychological sciences.

Lets consider the current status of the Military Industrial Complex and the PR themes that are being used to support it. As I argued earlier the size and capabilities of the U.S. military are far beyond what is needed to act as a deterrent of an invasion of the United States by a foreign power. If we consider the content of the current PR effort it has shifted from creating fears of communism to creating fears of terrorism. Much of the justification for the colossal defense expenditures is to have adequate means to fight terrorism. If the nature of terrorist activities are scrutinized, it is comprised of small numbers of people in scattered locations. These are not large scale military formations maintained by nation states for defense and/or conquest. The terrorist activities have several political objectives: the first purpose of their activities are to modify the national policies of their targets. The second is to garner support of sympathetic governments and their populations, basically it is an aggressive and destructive form of PR. If we recall the stated objectives for the recent invasion of Afghanistan by the United States, the invasion was justified by the US government's demand that the Afghan government produce the terrorist Osama Bin Laden for arrest by US authorities which didn't occur. After the invasion and conquest of Afghanistan took place it turned out that Mr. Bin Laden decided to leave and wasn't waiting there to be captured or killed. If we consider terrorist group characteristics, I would argue that combating terrorism is really more of a police style of activity rather than one requiring large scale military forces. Having another $13.5 billion aircraft carrier or greater numbers of stealth aircraft simply isn't useful for this application. Fostering greater international cooperation would be though, and perhaps if we practiced a less aggressive international policy it would take some of the wind out of the terrorists sails.

From the current perspective how have the fears of communism played out? China which was one of the greatest communist Bogey Men during the cold war has turned out to be our largest trading partner. Viet Nam is also a peaceful trading partner and the domino theory used to justify the war with Viet Nam simply didn't take place.[13]

Of course the United States is still being threatened by communist Cuba, and the sanctions are still being maintained. Don't you feel threatened by Cuba? Also, why is Cuban communism bad and Chinese communism now OK?

To sum up, it seems that what is really the threat is one to the bottom line of the military industrial complex, no

---

[13]   *The domino theory indicated that if Viet Nam became a communist state the entire region would like a row of dominoes progressively knocking succeeding adjacent dominoes over.*

Bogey Man no justification for huge defense expenditures and profits!

~~~~~~~~~~

Chapter 2
Tracks

To be or not to be that is the question?

Hamlet by William Shakespeare

In this chapter we will consider a number of societies and the most predominant factors that affected their viability. The fundamental reason for a societies collapse, is an inability of the members of the society to produce appropriate responses to changes occurring in it's environment.[14] The contributing factors that underlay the maladaptive responses are, rigidity arising from ignoring obvious problems, and flawed appraisal of the situation. The most common underlying motivation for this type of thinking is persistent ignorance, which arises from being excessively invested in maintaining the status quo and/or an ideology. The result, is an inability to realistically evaluate the relative importance of options.

The first set of case studies: these are comprised of societies that collapsed as a result of an inability to match the size of the population to the resource base.

---

[14] *Diamond considered some of the same societies presented here in his book Collapse. The factors that he considers to significantly affect the sustainability of a society are: environmental fragility, climate change, failure to adapt to environmental change, loss of essential trading partners and hostile neighbors. Pp 11-15*

Easter Island is located 3,500 km west of the coast of South America and is one of the most remote islands in the Pacific Ocean. It was originally colonized around 300 – 400 CE by Polynesians.[15] Using pollen analysis, tree casts, and plant remains, it is known that prior to the arrival of the Polynesian colonists the island was heavily wooded with tall trees and woody bushes, composed of 16 plant species, the majority of them being trees. The largest and most abundant trees were a now extinct species of palm Paschalococos Dispertie, and is closely related to the Chilean Wine Palm. The Chilean Wine Palm is still found in South America where these trees grow to be 19.8 meters (65 feet) tall with a trunk diameter of 1m (3 feet). The Paschalococos Palm trees found on Easter island were the largest type of palm tree, casts of trees that were buried in material from volcanic eruptions have been found that are as large as 2.13 meters (7 feet) in diameter. The sea surrounding the island had numerous fish species and significant populations of porpoises. The underwater profile of the base of the island slants down steeply causing the water to deepen quickly thereby, significantly limiting fishing directly from shore. The climate of the island is mild and receives adequate rainfall, enough to sustain agricultural food production. The surrounding ocean also could be used to provide additional food from off shore fishing. All in all, one could say that Easter Island was well suited for human occupation. The greatest disadvantages for a human population was its isolation which precluded oceanic trade and its small size (163.6 square kilometers).

When the Polynesians set out to colonize a new island they carried the kinds of plants and animals they already had

---

[15]   *This date is the generally accepted value but is not firm, the earliest radio carbon dating is from 700-800 CE.*

been cultivating or husbanding for food to be established in the new location. The kinds of food that the voyagers established on Easter island were chickens and the cultivars taro, bananas, paper mulberry and sugar cane. They usually were also accompanied by a few unintended additional species that also arrived by hitching a ride primarily, rats and insects.

Upon their arrival at Easter island the Polynesian pioneers established their settlements on the most favorable areas of the coast. The areas chosen were those that were most agriculturally productive and also where it was convenient for launching fishing expeditions. The pristine island provided a robust means of subsistence for the early settlers. Over time the population increased, in the initial stage of population expansion the remaining most desirable areas were exploited for food production. As the population continued to increase the less desirable slanted hillsides were cleared of forest and put into agricultural production. At the same time further pressure was placed upon the islands forest resources to provide fuel for cooking, and building materials in order to produce boats, dwellings, and equipment.

The islanders practiced ancestor worship which required the construction of colossal statues that were generally around 10 meters high and weighed 82 tonnes. The largest of the erected statues was an astonishing 18 meters high and a even larger one that was being sculpted was 21 meters in length and weighed 270 tonnes. The golden age of Easter Island statue construction occurred between 1200 – 1500 CE, producing 887 statues. In order to move the statues to their mounting pedestals and erect them required much large rope and many large timbers, thereby, placing further demands on the islands forest resources. Physical evidence suggests that deforestation of the island was complete some time between 1400 -1600 CE which

49

coincides with the end of the statue building era.
Moreover, in 1722 the first European contact was made by
a Dutch ship captained by master Roggeveen, who noted
that the island had no forests but many large standing
statues. Captain Roggeveen remained at the island for
only a short stay because he thought that the islanders
where cannibals. The island was subsequently visited in
1770 by a Spanish ship whose records indicated that the
statues were still standing. Later in 1774 Captain James
Cook visited the island and reported that many of the
statues had been toppled. The prevalence of cannibalism or
how long it may have been practiced is not known. Some
additional sparse physical evidence exists besides Captain
Roggeveen's suppositions, that indicate that cannibalism
was practiced. This evidence is comprised of tool marks
found on human skeletal remains that are indicative of their
use for the process of dismemberment. During this period
the Easter islanders where also living in stone huts with
very low entry ways that requires one to crawl in to the
dwelling. This type of architectural feature is usually
incorporated into buildings for defensive purposes. A low
entryway, places an intruder in a vulnerable position when
entering, further suggesting that violence was a feature of
the society at that time. The population when Captain
Roggeveen visited has been estimated to be around 2,000 –
3,000 persons, down from around 10,000 persons in the
15th century. Archaeological evidence suggests that the
population declined from the 15th century level gradually
to the 18th century value. In the 19th century it was further
reduced to around 1200, as a result of abductions by slavers
and the accompanying violence associated with this
practice. In 1868 slavery had been made illegal in South
America and the 15 surviving Easter Islanders were
repatriated at that time. By the 1870s the population of the
island had plummeted to just over 100 persons as a result of

the introduction of novel European diseases that the islanders had no immunity to. Since then the population has increased to 3,000.

The two most significant underlying factors that caused the population to decline from a height of around 10,000 to around 3,000 was population expansion beyond the carrying capacity of the island and an inability to import resources from remote locations. After the arrival of the Polynesians their population continued to increase requiring the conversion of the pristine land they initially found into agricultural production. Initially the low lying areas were put into food production until all the choicest parcels of land were being used. Since the location of Easter Island is so remote, and the amount of cargo sailing canoes can carry is small, it was not practical for them to import food to supplement domestic production. The remaining options for them to meet their needs, was to limit population growth or increase the level of exploitation of their immediate area.

They chose to increase the level of exploitation of the local resources available to them by clearing the forests that occupied the hillsides and the flanks of the volcano, thereby, increasing the area of agricultural production. This strategy did provide a temporarily expedient to meet the food needs of the expanding population. However, the flaw that made it a temporary expedient occurred as a result of the sloping topography of the newly exploited land. Erosion of sloping forested land is prevented by trees which immobilizes the soil through a number of processes. These processes originate from the roots of plants which physically hold onto the soil, allow rapid absorption of rain water, and reduce the flow rate of surface water. The forest canopy is also important in preserving good soil condition because it attenuates the force of the impact of the rain falling upon the ground and also slows down the

51

rate that rain reaches the ground. A further benefit of stands of timber are that they serve to reduce the effects of erosion caused by wind by reducing its velocity, and capturing dirt particles. A third consequence of reducing forest cover is that it reduces habitat for birds who provide significant amounts of nutrients to the soil from their droppings (guano). Thus, when the islanders cleared the land of trees it became rapidly degraded through the processes of erosion, loss of nutrient sources, and reduced water retention. The combination of these factors resulted in a falling off of productivity. A further consequence of hillside soil erosion is that it moves the soil in a downward direction and ends up in streams which transport it to coastal areas. Once the silt arrives at the coast it settles out covering the sea floor with mud. The coating of mud on the sea floor has the effect of reducing the productivity of inshore aquatic resources.

As the process of soil erosion and degradation continued, it became necessary for the islanders to continuously move up the hillsides to replace falling food production. The food shortfalls were exasperated by the continuing expansion of the population. Ultimately the forests were completely cleared, thereby denying them building materials which eliminated boat building and terminated statue production. After boat building ceased off shore fishing was no longer feasible, further diminishing their food resources.

We know from archaeological evidence and population estimates from European travelers that the population was around 3,000 in the 18th century, and that it did plummet from around 10,000 over a period of several hundred years. It is obvious that they did not choose to voluntarily limit or reduce their population to match their diminishing resource base. This can be surmised from evidence that a change took place in their diets to less desirable types of food such as rats,[16] which increase to 60% of their animal based food while their other sources of food diminished.

Further support for this hypothesis is the fact that defensive features were incorporated in their buildings to provide protection from their fellow islanders[17] to forestall food theft and cannibalism.

Classical Maya: The classical Mayan civilization occupied an area in meso America that is now comprised of parts of Chiapas and the Yucatan peninsula in Mexico as well as in Guatemala, Honduras, Belize, and El Salvador, an area of 324,000 square kilometers. The sea coasts touching The Gulf of Mexico, Caribbean Sea, Gulf of Honduras, and the Pacific Ocean had only a few major population centers (kingdoms), the majority were located located in the interior. This area had been occupied for a long period before the formation of the classical style of society which existed from 250 CE through 1000 CE. The

---

[16]   *Rats are generally not favored as a food source. In the Paris commune that replaced the second empire in 1870 rats were used for food because the of the siege, they were quickly abandoned once other food sources became available. The same thing occurred in world war II prisoner of war camps. None of these people continued to eat rats when other food sources became available. Rats are generally not favored as a food source. In the Paris commune that replaced the second empire in 1870 rats were used for food because the of the siege, they were quickly abandoned once other food sources became available. The same thing occurred in world war II prisoner of war camps. None of these people continued to eat rats when other food sources became available*

[17]   *This style of dwelling was being constructed before contact with outsiders occurred precluding this possibility for a motive to use this type of defensive entryway.*

classical type society didn't suddenly appear but was an outgrowth of prior societies that already incorporated many of the defining traits. Because of the large area this civilization occupied the topography and climate are fairly varied, and therefor the historical trajectories the various kingdoms followed are also varied. The main focus will be on the group of kingdoms that are referred to as the Southern Lowland Maya since they have been the most comprehensively studied.

The Classic Maya were capable of working the metals copper and gold, had a system of writing, a sophisticated understanding of mathematics that included the independent invention of zero, and precise calendars based upon astronomy which also embodied their mathematics. They did not have draft animals or use the wheel for daily activities. Wheels were known but only used on a few religious statues to move them around. Thus, transportation of commodities was accomplished by people carrying things. Foot transport acted to limit trade to smaller carry-able items such as obsidian for tools, materials to produce religious artifacts, and prestige items for the elites, some of which came from long distances. Because of the transportation limitation the size of the kingdoms were small. Generally the Maya "cities" were comprised of a central core area consisting of large monumental buildings used for religious purposes, palaces for royals and nobles. They also had ball courts which were important for religious activities as well as sport.

The religion of the Maya was composed of two divisions incorporating three elements, the heavens and underworld where gods resided, and the earth which was occupied by humans. Blood letting rituals were conducted at the temples by the royals to achieve visions that were believed to originate from their ancestors or the gods. The visions were induced by a combination of bloodletting and hallucinogens derived from the Bufo Marinus frog. These visions were thought to be conveyed to the recipient by underworld serpents.

The ball game[18] reflected the game of life and demonstrated peoples place in the cosmic order. The game was associated with military activities that were endemic to the Mayan kingdoms. It ritually reenacted successful military campaigns and established the connection between the earthly plane and the gods who required blood sacrifices. The blood sacrifices were thought to resurrect the gods who followed the cycles of the sun, moon, and stars. These cycles were also used to predict planting times. Each of the gods would undergo periodic death which was synchronized with the cyclic intervals of the natural process the god embodied. The royals when they died were resurrected from the underworld as a god. The most desirable of the pantheon of gods for royal resurrection was the corn god (over 60% of their diet was corn). The corn god symbolized planting corn on earth and like the plant suffers cyclic decapitation (harvesting). War captives where used to play the ball game against their conquerors for ritual purposes. The ritual ball games outcome was never in doubt though, the captives always lost and were subsequently beheaded or had their hearts ripped out to provide a blood sacrifice. The captives were also utilized for sacrifices in numerous other religious

---

[18]   *The ball game was also played for non ritual or sporting purposes and both men and women were participants. It used a heavy rubber solid ball and is the precursor of modern ball games using rubber balls such as basketball,soccer, volleyball etc. The game was played by hitting the ball with hips thighs and upper arms but not kicked or thrown and passed through hoops mounted on the walls, the players wore protective clothing.*

rituals connected to the various gods to revitalize their presence. It was a common practice for the victims to be tortured and mutilated for protracted periods before being sacrificed. Thus, the well being of the kingdom was dependent upon the king because he was a conduit to the gods since he was the author of the constant blood sacrifices. These sacrifices were of several types. The first was conducted by the royal who cut themselves and dribbled some blood on a piece of paper which was burnt to convey it to the gods. The second method used sacrificial animals and people by killing them in various ways.

In the earlier part of Mayan history, swidden (slash and burn) a dispersed form of agriculture was practiced. The farmer practicing this method would clear a patch of forest and burn the cleared vegetation which provided ash to stimulate growth. However, tropical soils are thin, most of the nutrients are concentrated in the growing vegetation and the soil becomes depleted in 2 to 4 years when used for this type of agriculture. After the soil becomes depleted and crop production falls off, the farmer would move to a new location and repeat the process. The depleted field would be allowed to lie fallow for 10 to 20 years to recover its fertility and then returned to for crop production. This form of agriculture does not require a great deal of labor, and can be practiced with simple hand tools. In the 1930's studies were conducted on swidden agriculture. The findings of these studies showed that it only required 48 days of work to produce enough food to feed a family. This method of agriculture can be sustainable as long as adequate periods of fallow time are allowed for the soil to recover its fertility. Because of the long fallow periods it requires a large amount of land to practice, and therefore can only sustain low population densities.[19]

In addition to the cultivation of plants, Maya archaeological sites have provided evidence of hunting and gathering of wild foods found in the surrounding forests and aquatic environment. Appraisals of the quantities of food remains indicate that their diet consisted principally of corn, beans and squash, supplemented by minor amounts of avocado, coca, palm fruits, and dogs. The favored wild foods were white tailed dear, peccary, fish, mollusks, and wild forest mammae apples.[xii] Because of the large geographic area

---

[19]   *The Kayapo a modern South American Indian tribe practices sustainable swidden agriculture that mimics the process of ecological succession. They cut a patch of forest at the start of the dry season in April-May and leave the plant materials to dry until astronomical observations and a number of biological indicators such as the flowering of indicator plants and animal migrations start to take place in late August – September and set on fire when the meteorological conditions are correct.. The process of burning is carefully supervised by their shamans being focused on the combustion of small trunks and lianas but kept under careful control. Several weeks prior to the burning the soil is seeded with sweet potato, manioc, and yam which sprout immediately after the burn. Short cycle crops such as corn, beans , melon, and squash follow which quickly provide ground cover in conjunction with long cycle crops that can be harvested after 6 months to 2 years after planting. Just prior to "abandoning" the depleted field that will lie fallow they plant up to 16 species of trees and shrubs that provide desirable fruit. Returning to the plot occasionally to prune and do light weeding. The fruits of these trees are also consumed by game animals and birds which are attracted to the fallowed area providing an additional source of wild food. Moreover, the game animals also establish additional fruit bearing plants by seeds carried in their droppings further enhancing the plots as a resource while providing an essential part of the recuperative process. From: The Fate of the Forest by Hecht & Cockburn.   pp. 45-50.*

through which these kingdoms were located, the level of consumption of the various food varieties reflected local availability. For example, kingdoms that had access to aquatic resources incorporated more of these food types in their diets. To get a sense of the amount of variation among Maya kingdoms a few will be used as examples. We will start with Copan located in the interior of Honduras. The rise and fall of Copan's population was worked out by Ann Corrine Freter of Ohio University using obsidian hydration dating, and Wingard who assessed the carrying capacity of the area. We will start in the mid to late 6th century CE when all of the alluvial soil was in agricultural production. At this time Wingard estimated that around 5,000 people resided in this area.

Date CE......................Populations
600– 650........................5,520
650-699........................10,426
Population Exceeds Alluvial Carrying Capacity
700-749........................15,009
750-799........................27,773
Population Maximum
800-849........................25,656
Royal Dynasty gone.
850-899........................26,381
900-949........................14,821
950-999........................12,118
1000-1049...................7,698
1050-1100....................3.813

After the population exceeded the alluvial carrying capacity, the less desirable forested lands on the slopes were exploited to provide sustenance to the burgeoning population, thereby causing progressive environmental degradation from severe erosion and soil depletion.[20] By the 8th century deforestation was probably complete. The population continued to grow, and became progressively more impoverished as a result of an inability to produce adequate amounts of food.

Archaeological research suggests that the farmers did not have permanent rights to a parcel of land. Land rights where possessed by the royals and controlled through the nobles who would allow a farmer the use of a parcel on a temporary basis. There is no evidence that the elites ever initiated any measures to address the declining agricultural production. This inactivity is puzzling since they did have a knowledge of terracing. Terracing uses retention walls to immobilize soil and is an effective means of preventing erosion on slanted fields. Perhaps an explanation can be found for this failure by looking at other more documented societies such as 19th century Ireland, where the elites were also "absentee landlords." Typically absentee landlords lack involvement in the actual work taking place. Because of their detachment from the property under their control they seldom had little or no idea of what was actually occurring. Generally if elites start to feel a little "pinched" their remedy is to intensify extraction, and carry on as before. Moreover, the people working the land as "tenants"

---

[20]   *David Webster excavated a Maya building in this area that had walls that are 7 meters (23 feet) high which were nearly covered by eroded soil. The top layer of newly formed soil was only 4 – 8 cm thick and mostly comprised of roots after approximately 700 years regeneration.*

also have no incentive to invest a lot of labor to install and maintain improvements like terraces, since they don't derive any immediate or ongoing benefits from their labor. Since terraces require substantial amounts of ongoing maintenance, protracted attention from the people in control is essential. Needless to say, a sustained focus on these kinds of mundane details are not a hallmark of absentee land lords.

According to the above table, the population of Copan reached a maximum in the late 8th century whereupon the dynasty fell. The last King in the direct royal lineage was Yax Pasaj who acceded to the throne in 763 CE and is last heard of in 810 CE. A final reference to a ruler named Ukit Took was inscribed in 822 CE on a partially finished alter.[xiii] He appears to have been an unsuccessful pretender. This is the last entry that is relevant to any royal presence in Copan's written record.

The termination of the royal presence in Copan is not accompanied by signs of strife. In many of the other Mayan kingdoms there is evidence of strife and that the royals were forcefully removed. The deposition of royals probably occurred as a result of the obvious ineffectiveness of their system of religious beliefs to address their desperate situation. This impotence probably led to a general disillusionment among the populace and an abandonment of their religious creed.

As can be seen from the above table in the early 10th century the population of Copan experienced a large population reduction that was followed by a continuing decline at a gradually slowing rate. It's believed that some of the elites were able to persist for some time after the end of the classic period, because they still provided some useful services to the moribund society. By the time of the Spanish conquest in the early 16th century, the area was completely abandoned as reported by Cortes who passed through the area and nearly starved.

The Maya Kingdom of Lamanai was located on the shore of a lagoon of the New River in Belize. As a result of the location of this settlement, it had a more diverse resource base that incorporated locally obtained aquatic fresh water foods comprised of fish, shellfish, and turtles. In addition to freshwater fish, there also is evidence that seafood was a part of their diet. It is thought that the seafood was transported by boat on the New River from the Atlantic coastal areas where a maritime trade network was in place. The maritime trade network was large, it extended along the Caribbean costs of Mexico and Belize. The most prominent of the oceanic trading settlements were Jain Island, Frenchman's Kay, Moho Cay, and Wild Cane Cay.[21] Wild Cane Cay occupied a position as the last link of the coastal network. It was also the start of the New River connection to the inland kingdom of Lamanai. All of these settlements survived the classical Maya collapse and persisted into post classic times.

I would hypothesize that the Lamanai kingdom and maritime trading settlements were able to weather the classic collapse as the result of the availability of external sources of food. These more diverse food sources were provided by a more capable transportation system based upon boats. Boats were used to ship marine resources inland by river while also providing access to the contiguous river waters. This larger subsistence base reduced the degree of dependency on locally derived

---

[21] *The coastal trading centers were comprised of traders and did not have large central monumental types of architecture associated with the kings. For example, Wild Cane Cay was densely populated but only occupied an area of 10 acres.*

agricultural food. Thereby, reducing the sensitivity of Lamanai to agricultural food shortfalls allowing it to carry on.

North America: During the initial establishment of the American colonies a nonexistent to rudimentary production capability existed. As a result of this deficiency, the early colonists were dependent upon trade with the parent European countries to provide manufactured items.[22] In order to pay for these necessities the colonists had to produce trade items that were of high enough value in Europe to produce a significant net profit. The profitable high value trade items that were obtainable by the early colonists were wild animal skins (fur) obtained from the hinterlands, and tobacco which was cultivated on farms and plantations. Most of the fur was obtained from the Indians through trade by the French and English colonists who had established porous territorial trade networks with the Indian tribes. The preeminent source of wild animal products was the Iroquois confederation, which was comprised of five tribes: The Mohawk, Oneida, Onondaga, Cayuga, and the Seneca.[23] This confederation was also by far the most

---

[22]   *The British colonial policy with respect to trade was based upon mercantilism which operated by the transfer of raw materials from the colony to the mother country where these materials where manufactured into finished products. It was illegal to transfer manufacturing technology to the colonies to maintain colonial dependency on the mother country and also to have a captive market for English products to produce profits for the English industrialists.*

[23]   *The five tribes were the primary members of the confederation and undertook diplomatic endeavors for themselves as well as other tribes such as the Cherokee who were referred to as "props" and did not have their own negotiators.*

powerful group of Indians, it is estimated that their population was around 80,000 in the British sector and 20,000 in the French sector in what is now Canada. They had highly skilled "professional" diplomats referred to as sachems who were appointed for life giving them high levels of experience which resulted in great political savvy and negotiating skill. Their diplomatic strategy was to play the French and British off against each other, and it was quite successful until the American revolutionary war disrupted the balance of power.

The monoculture cultivation of tobacco is one of the most taxing crops on the soil to grow, it was also the most lucrative crop. Because of its high returns, the colonial farmers grew as much of it for a cash crop as possible. The colonists farming practice was to plant it time after time until the soil became too depleted to produce significant yields. Once the yields became too low, the colonial farmers would abandon the depleted croplands and relocate in an area with virgin soil. This pattern of land usage was practical because land prices were very low. It has been argued by Montgomery in his excellent book Dirt, that the loss of soil fertility and the resulting decline in crop yields in the south Atlantic colonies plantations were a significant underlying driving force for the adoption of slavery in order to reduce labor costs, thereby enabling the planters to continue to produce a profit from their property.

By the 18th century soil depletion and population expansion was creating a strong incentive among the citizens of the British colonies to expand westward past the Appalachian mountains to acquire more land. There was an impediment to this expansion though. The Iroquois confederation had negotiated treaties with the various colonial governments operating under the authority of land grants made by the British crown. These treaties provided the basis for the Iroquois to regulate trade. The method the

Iroquois preferred for this purpose was to restrict commercial access to enclaves in their domains. The colonists operating on the principle that they were here and the British government was far away, ignored the treaties and expanded into Indian territory. The members of the Iroquois confederation became vexed by the blatant violations of the treaties. To rectify this situation four sachems were dispatched to England to lodge a protest against the violations. In England they were granted an audience with queen Anne for discussion of the problem. Later a treaty addressing their complaints was embodied in the Royal Proclamation of 1763 forbidding any settlements west of the Appalachian Mountains. However, the colonists continued to encroach upon the Indian territories and further treaties ceded land in West Virginia, Kentucky, and ended with the treaty of Lochabar in 1770 that relinquished Cherokee land that extended as far as Kingsport Tennessee. Even with these large transfers of land the colonists continued their land grabs. The land grabs were given a bogus appearance of legality by establishing unauthorized land offices in Indian country that were used to produce illegal titles to transfer ownership of these lands. Many of the land speculators engaged in these activities were the colonial elites who later agitated for independence from Britain and also participated in the revolutionary government. After the war of independence the new American government recognized these titles making their holders owners of vast tracts of land.

The social climate essential to support slavery was changing though. In the late 18th and early 19th century an efflorescence of religious feeling which condemned the practice of slavery took place in Britain resulting in it's practice being declared illegal in 1807. Shortly thereafter, the British government dispatched naval forces to interdict the shipping of slaves from Africa, and to be an instrument of condign punishment to the shippers. This resulted in a large decrease in the number of new slaves being landed in the United States and an increase in their value (this will be returned to shortly).

After freedom was gained from Britain, the United States entered a period of vigorous expansion, no longer being hindered by British treaties with the Indians. As these new territories were occupied by settlers from the original colonies, they soon wished to change their status from being a territory to that of a state to obtain political representation and other benefits from the federal government. These applications to the federal government for statehood became a bone of contention that eventually culminated in the civil war. The conflict occurred because southern plantation states wanted slavery to be a legal practice in new states, while the northern non-slave states were adamantly opposed to the expansion of slavery. The question is why was it so important to the southern states to extend the practice of slavery? In Soil, Montgomery's analysis of the economies of the southern slave states showed that the progressive depletion and loss of fertility of their soil by the first half of the 19th century had reduced crop yields to such a degree that breeding slaves had become a major industry. The importance of this industry was greatest in Virginia, Maryland, the Carolinas, and Georgia. Without the economic prop of slave transactions in these states the plantation system would have collapsed. For example, in the 1850's, Georgia's slave breeding business was the largest source of income. Slave industry estimates at that time indicated that if Missouri, Texas, and California were to become slave states the value of slaves would double while also providing vast new markets.[xiv]

The same pattern of feckless land exploitation has continued, in 1930 a drought that lasted till 1940 occurred in the southern great plains which is referred to as the Dust Bowl. It was characterized by numerous intense episodes of dust storms. The dust in these storms originated from the rapid and severe wind erosion of the soil. The affected area was 39.25 million hectares (97 million acres) encompassing southeast Colorado, Northeast New Mexico, west Kansas, the panhandles of Oklahoma and Texas. It is a roughly potato shaped area having a north – south axis of 644 km (400 mi.) and 482 km (300 mi.) on the east – west axis. The dust storms took place during the windy part of the year in February, March, and April. They did not occur over the entire area at one time. The first severe storm occurred in 1931, and the peak of the entire series of storms occurred in the mid 1930's. In 1938 the amounts of precipitation started to increase which produced a tapering off of the storms, by 1940 the episode was over.

Agricultural development of the dust storm area was initiated and fostered by the US government. The instrument used was the 1862 Homestead Act which was passed by the US congress to encourage settlement in the Dust Bowl area. This act was formulated and implemented without consideration of the conditions prevailing in the area. During the period from 1862 through 1930 a number of severe droughts occurred the first from 1860 – 1864, another in 1910 – 1918. All of these droughts were accompanied by severe wind erosion events that gave rise to large dust storms. Having experienced a number of these drought induced dust storms, the inhabitants of this area were well aware of these types weather events. They also were cognizant that cultivated land using traditional plowing methods were prone to erosion.[24]

---

[24] *More recent studies have shown that droughts were a periodic occurrence in this area, after the last ice age dendrochronology (tree ring studies) indicate that 21 major*

Starting in the mid 19th century great strides were made in agricultural technology that enabled progressively more land to be worked by a farmer through innovations in plowing and harvesting machinery. In the early part of this period a number of draft animal powered implement improvements appeared such as the sulky plow which allowed the farmer to ride on it, and the Lister plow that combined several operations, incorporating a seed hopper and seed drilling device that operated during plowing. These types of innovations greatly increased the area that could be plowed and seeded per day. In addition to improved planting implements, harvesting innovations that combined multiple operations also appeared, such as the header which cut only the heads of the grain and left the straw. By 1885 steam traction engines were in wide use, their greater power and versatility allowed the use of multiple furrow gang plows and powered threshing machinery. The combination of these technical innovations greatly augmented production capability. Ultimately internal combustion equipment appeared producing even greater capacity for tillage and planting. At the same time the combine appeared which is an automated system that incorporates all the harvesting operations.

More recent studies have shown that droughts were a periodic occurrence in this area, after the last ice age dendrochronology (tree ring studies) indicate that 21 major droughts that lasted at least 5 years had occurred in this area. From 1210 CE – 1958 CE the average drought persisted for 12.8 years and droughts that lasted for at least

---

droughts that lasted at least 5 years had occurred in this area. From 1210 CE – 1958 CE the average drought persisted for 12.8 years and droughts that lasted for at least 10 years occurred every 55 years.

10 years occurred every 55 years. The result of all this increased capability was the expansion of farm size, for example from 1910 to 1920 farm size increased from an average of 188.4 ha (465. 5 acres) to 312.4 ha (771.4 acres). To accommodate the increasing farm sizes much virgin land had to be put into production going from 4.05 million ha (10 million acres) to 7.128 million ha (17.6 million acres) in the same period (1910 – 20). Much of this newly exploited land was submarginal for growing crops. All of the newly converted virgin land in this area was comprised of native grasses that shielded the land from erosive forces. In addition, the number of cattle increased from 506,583 to 894.859 head (Hurt, p. 23) which further reduced the grass cover by grazing. In tandem with the agricultural expansion, the transportation system was greatly enlarged and improved. By integrating the dust bowl area into the wider national transportation network, access was enhanced which brought in greater numbers of newcomers. The expansion of transportation also augmented the available market area introducing a further driver for increasing exploitation.

The most significant underlying factors contributing to the dust storms, was the use of unsound traditional plowing methods combined with periods of exposure of bare soil, which amplified the effects of drought and high winds. No techniques to retain soil moisture were practiced by the farmers[25], instead they allowed the ground to be exposed by the practice of burning off crop stubble. Stubble provides the organic materials that produce humus. Humus is an essential component of soil structure because it augments

---

[25] *The average precipitation in this area is 45.7 cm (18 in) which is just on the margin required to produce the kinds of crops that were commonly grown.*

the soils capability to absorb and retain moisture. After 5 – 7 years of burning stubble and plowing, the soil was completely pulverized seriously degrading its fertility. Pulverizing it to a fine powder also increases the ability for the wind to lift it and carry it off. The results of these unsound planting practices were aggravated by further reductions of stubble and the formation of bare ground which was produced by the practice of grazing excessively large herds of animals in the fields.

The crops that were the primary focus of production were wheat, corn, and cotton , wheat being the most favored. These crops did well during periods when the amount of precipitation was high. However, during periods of low rainfall they did not grow well and they were also susceptible to damage from blown dust. After the onset of the drought cycle the farmers persisted in planting these crops, gambling that adequate precipitation would appear. These gambles were taken because these were the most lucrative crops. The lower value more drought resistant plants such as sorghum, sudan grass, and small grains were not widely planted.

Another source of the poor agricultural practices originated from the fact that many of the farms in this area were not worked by the owners. The owners were often absentee land lords or suit case farmers with other sources of income in town. The tenant farmers and share croppers that actually worked the land were interested in maximizing potential yearly earnings. Little incentive existed for them to grow lower value drought resistant crops that reduced soil erosion, or invest their time in other measures of soil conservation. When you see pictures of the "Okies" clearing out for California, these are almost all tenants or share croppers who were let go by the owners. Few of the owners actually lost or were forced to sell their land.

The owners were able to retain their property because of numerous government programs that provided various kinds of assistance. Large numbers of cattle were bought up by the federal government and slaughtered to thin the excessively large herds and provide a financial boost.

The meat was canned in 100# and 50# tins for use in various other programs, such as school feeding and military stockpiles. Train loads of hay and other animal feeds were shipped in at nominal prices to support the remaining herds. Government programs that provided subsidies and/or labor to install terraces, Lister plowing,[26] and shelter belts,[27] were widely available. Shelter belts were wind breaks comprised of shrubs and trees that were installed across fields and around farmstead buildings to reduce wind speeds and soil drifting. A number of additional restorative strategies and subsidies were implemented by the federal

---

[26] *Lister plowing uses two plow blades that roll the soil towards each other creating a central raised ridge bordered with furrows. The rough ground surface reduced wind velocity allowing blown soil to be captured in the furrows as well as producing a "sink" that captured moisture and reduced evaporation losses because the moisture was sheltered from wind.*

[27] *The shelter belt extended from Bismark North Dakota to Amarillo Texas the western boundary was set at the 40.6 cm (16") precipitation line and the eastern boundary at the 55.9 cm (22") rainfall line. Each segment of the belt was 132' wide and spaced at 1 mile increments oriented in an east – west direction. It utilized red cedar, hack-berry, and green ash in the center which was the highest portion with a number of progressively lower growing shrubs out towards the edges to lift the wind. 18,600 miles were planted containing 217,378,352 trees, the survival rate of trees varied, the average was 58%. The trees produced a wind shadow that extended out 20 times their height.*

and state governments such as strip cropping,[28] grazing restrictions, and prescribed care for shelter belt vegetation. Further government subsidies were granted to underwrite the use of drought and wind resistant crops, in addition to replanting grass lands. All of these measures were reinforced by financial assistance programs.

In 1938 the drought began to weaken and by 1940 the rain fall was up to the levels that were typical of the wet part of the climatic cycle. The advent of World War II created a strong and continuing demand for food, producing robust prices for agricultural commodities. The farmers and ranchers were quick to revert to their old habits of over expansion of livestock herds, bringing submarginal lands into cultivation, and a focus on the same three crops as before, in order to maximize and augment short term profits. The younger farmers, tenants, and share croppers also frequently removed the shelter belts and abandoned Lister plowing, strip cropping, restricted grazing and other conservation measures. The older farmers tended to retain many of the conservation practices that were deployed in the 1930's. In 1950 a new drought started and persisted through 1955. The effects of the drought upon the agricultural community was similar to that in the 1930's. The government restarted many of the old programs, buying up large quantities of cattle and provided many of the same kinds of assistance as already described above. The effects of the drought on the agricultural community was not nearly as sever though. The impact of the 1950's drought was attenuated because the government acted quickly to rapidly implement the techniques that were

---

[28]   *Strip cropping is comprised of alternating contoured crops composed of soil holding crops and densely growing feed crops.*

developed in the 1930's. Also the absence of the debilitating effects of the great depression were not present.

Japan: In 1603 CE the Japanese shogunate was established by Tokugawa Ieyasu the supreme general at that time, and continued until Tokugawa Yoshinobu returned political authority to the emperor in 1867. This period of Japanese history is referred to as the Edo period (Edo was the original name of Tokyo) where the showguns were headquartered. During this 265 year period the Japanese operated a sustainable society.

The sustainable society was achieved by utilizing things to their fullest extent. Durable goods such as metal pots, pans, ceramics, and barrels were repaired. Not only were durable items repaired but even such ephemeral objects as clothing, umbrellas, footwear, etc., were maintained by a comprehensive array of specialist repair people. Agricultural amendments and fertilizers were also obtained from waste products. These substances were produced by collecting the ash produced by fuel wood and human waste (night soil), which were used in combination to produce an excellent fertilizer.[29] Night soil in urban areas was an

---

[29]   *In the United States many municipalities produce compost whose principal ingredient is sewage for application to non food producing plants. If night soil is composted using the hot heap method which achieves temperatures of 50 C (122 F) for 24 hours the finished compost is sterile (Jenkins p. 152). Unfortunately municipal sewage contains industrial and other contaminants that usually make it undesirable for crops. Recently I was chatting with a landscaper who told me it was his "secret ingredient" for growing really lush plants. My uncle and aunt used to use composted humanure and they had fantastic gardens, this was back in the 1950's so this is nothing new but definitely an underutilized resource. In Japan the use of night soil for crop fertilizer is still a wide spread practice. It seems unlikely that it is the origin of diseases since the Japanese enjoy the longest life expectancy and best health in the world, perhaps, we could learn about this from*

important source of income for landlords who had contracts with farmers and/or dealers. In fact there are historical reports of friction between landlords and tenets about the owner ship of night soil because of its high value. Thus, the consumers of the crops were the producers of the fertilizers for the crops. This system automatically adjusted fertilizer production to match food requirements since it was regulated by population size. For example, if the population expanded it produced proportionately increasing amounts of fertilizer used to produce more food, thereby, establishing a closed appropriately scaled self sustaining system.[xv]

The other important factor for sustainability was that the population remained relatively constant, its level did not exceed the carrying capacity of the environment. The level of population is known because the Japanese government started taking population census in 1720. This initial census found that the nation was comprised of 30 million people. In order to achieve population stability the Japanese people employed a number of strategies, they married later, nursed babies longer, practiced abortion and infanticide as well as contraception. As a result of these measures the population remained constant until the 1867 Meiji restoration. Part of the Meiji government reforms were comprised of the establishment of a capitalist economic system based upon the western model of extraction and expansion.

Between 600 CE – 850 CE Japanese forestry practices were on a path of exploitation similar to those that took place on Easter Island. Totman identified the driving forces and

---

them it would definitely have an impact on the amount of chemical fertilizers being used whose manufacture produces large amounts of pollutants as well as diminishing soil tilth.

status of forest usage as originating from the "demands by the ruling elite for timber, to supply armies and build castles and religious monuments, had caused serious deforestation."[xvi] The pattern that was followed was to severely deforest an area and abandon it, then move to a new area which would be exploited in the same way. The village populations welcomed this activity because it cleared land that they would then use for agricultural purposes. This practice worked in the initial stages of this period while enough forested land remained to provide their needs for agricultural supplements and other forest derived products.[30]

In the medieval period the Japanese population was sparse, by 1570 Japan's population is estimated to have attained around 10 million. This population expansion proportionally increased the demand for subsistence forest products.[xvii] By around 1670 the deforestation had become pervasive and was recognized as a problem. Up until the end of the Edo period, there was no land ownership in the sense thought of in western societies where one gains title to a parcel. It operated by a system of tenure. Land ownership was prohibited in spirit by the Tokugawa Shogunate, it was structured as follows: the Daimyo (a lord) had tenure in his domain, communal tenure was the next level, and the final level was individual tenure. These land usage rights were granted on a long term basis which provided motivation to conserve and improve these resources. Almost all Japanese forested land was Daimyo tenure or communal tenure.[xviii] In this system the farmers

---

[30]   *The forest was a source of clean water for rice field irrigation, household use, fuel wood and charcoal. Forest leaf litter and organic material were composted to produce fertilizer, and used for fodder. It required about 10 ha to support 1 ha of farmland.*

owned the things they produced which were comprised of agricultural commodities, wood, and cottage industry products. They were free to use or exchange these products after taxes were deducted. The underlying driver for the development of silviculture occurred in the 17th century. It was fueled by the rapid growth of urban centers which continued unabated throughout the Edo period and into post Edo times. The construction of the urban areas required large quantities of wood. They were also subject to periodic fires which produced additional wood requirements. All of the materials were domestically produced as a result of the isolationist policy of the shogunate which precluded imports, thereby eliminating augmentation from off shore sources.[31] To meet these

---

[31]   *In 1543 the Portuguese established contact and trade with Japan and 6 years later Christian missionaries arrived and started to proselytize and produced some converts. The Japanese gave the westerners a hospitable reception. However the Christian doctrine which demanded complete loyalty to a jealous god (1st commandment) created suspicion in the shogunate. It was also noted that the trade taking place was closely associated with Christianity, the Portuguese traders would only drop anchor in ports where the Daimyo would accept Christianity and eventually was considered by the Shogunate as a pernicious doctrine that taught people to contravene government regulations. In 1578 Hideyoshi promulgated a decree instructing all Jesuit missionaries to leave within 20 days, he did not enforce the decree immediately giving them additional time to go. The Roman Catholics did not go and persisted in proselytizing in 1614 Ieyasu again ordered them to leave and they still persisted their activities under cover. The Shogunate started to persecute the Christians 1622 and eventually between 1633 and 1639 five seclusion edicts were issued that totally proscribed Christianity, banned Portuguese ships from entering Japanese harbors, and forbade travel abroad. The 1641 edict confined the Dutch to Dejima an artificial island in Nagasaki Bay and allowed only the Dutch and Chinese to trade in Japan.*

needs tree plantations principally of the conifers hinoki and sugi (Japanese Cedar, Cryptomeria japonica) were established. Trees initially were planted by villagers on a part time basis, and later larger plantings were also produced by wealthy local merchants using local labor. It took until 1920 to completely restore Japanese forests. In Diamonds excellent book Collapse he identified some agricultural factors that supported the Edo period ecological recovery as follows: The agricultural advantages the islands possessed were young volcanic soil that was rich in minerals that allowed a rapid recovery of fertility in addition to high rainfall that supported rapid plant growth. These natural advantages were not nullified by excessive grazing. There were no goats, sheep, and a low numbers of horses all of which denude the land of plant cover.

The number of draft animals were matched to the amount of available fodder, thereby eliminating another source of overgrazing. All of these factors combined to prevent erosion and promote soil recovery. A well developed maritime fishing fleet produced an abundance of seafood which reduced the pressure on land based food production. The Japanese society also possesses to a marked degree the ability to evaluate, formulate, and adapt appropriate strategies to changing circumstances. These traits can be seen in their willingness to invest in the future, avoid squandering resources on warfare, diversifying their energy base by shifting from the exclusive use of wood for fuel to include coal.[xix] The pressures on their forests were further diminished by developing innovations in conserving energy. For example the hibachi, a small portable multipurpose stove was invented, it had the ability to be easily moved to places of need. By having portability this type of stove enabled them to heat only the room that was being used, instead of the entire building.[32] It also provided the further advantage that it could be used for cooking.

By the beginning of the 19th century the Tokugawa Shogunate was starting to become unstable as a result of pressures produced by urbanization and the expansion and refinement of economic activities. These changes shifted formal power away from the samurai bureaucrats who where implementing a rigid and obsolete government system. The Shogunate also became progressively more stingy in providing the stipends to the samurai impoverishing them. Many persons of the samurai class as a result of their pecuniary condition ventured into commercial activities either as a part time financial reinforcement or to a total commitment to commercial activities.

At the same time as these destabilizing forces were developing in Japan, the European nations and America had developed a bellicose attitude as a result of their greatly enhanced power provided by the industrial revolution. In America ideas of manifest destiny and of cultural superiority prevailed. These ideas were fostered and fueled by the easy victories over the Indians and Mexicans that were provided by the new technological superiority. The vast resources of Asia were viewed as a new area that could be exploited by implementing these same types of "colonial" practices utilizing the new military technologies. China in particular was being forced into granting highly unfavorable trade concessions that were accompanied by loss of authority in their own territory, in the form of foreign enclaves such as the treaty port system. The treaty

---

[32]   . *In the United States we call this zoned heating and consider it to be a modern innovation. The Japanese have been using it for hundreds of years!*

port system was established as a result of the opium wars (1839 and 1842) fought between Britain and the Chinese government who was attempting to halt the British traffic in narcotics. The treaty ceded Hong Kong to the British and established five Chinese Entrepots which were designated as open cities where "British subjects could reside, trade, and enjoy the privilege of extraterritoriality – that is , the right to conduct themselves in accordance with British laws and be tried by British consular courts, not Chinese judges."[xx] The other western nations quickly obtained similar agreements with a most favored nation clause that automatically granted any new concessions gained by a western nation to the rest. These areas were essentially governed by the Europeans and Americans as quasi-colonies. The Japanese where aware of what was taking place through their trade with China and also by Dutch accounts. The events taking place in China caused much Japanese concern. In 1844 King William of Holland wrote to the Shogun saying that "disasters now threaten the Japanese Empire."[xxi] He went on to say that any "nation preferring to remain in isolation at this time of increasing relationships could not avoid hostility with many others." The Japanese officials declined an offer included in William I I's letter to aid the Japanese to become integrated in the new globalized system of commerce.

In America, the business community saw Japan as another Asian nation that could be exploited for the extraction of profits and at the same time the Christians viewed Japan as an area for missionary activities. These groups were actively lobbying the government with the result that president Millard Fillmore sent a naval squadron under the command of Commodore Matthew C. Perry which arrived at Edo Bay on 8 July 1853 to carry out these objectives. Six days later Perry went ashore with much military fanfare and delivered Fillmore's and several of his own letters to the representatives of the Shogunate which stated that Japans policy was "unwise and impracticable" and urged them to avoid an "unfriendly collision between the two nations, by responding favorably to the propositions of amity, which are now made in all sincerity." Perry indicated that he was departing forthwith and would return "with a much larger force" if necessary in the following spring for the Shogunate's response. The Japanese had been strengthening their coastal defenses, but knew that they were far from adequate to fend off a potential belligerent response if the American demands were ignored. On 14 February 1854 Perry was back with an even larger contingent of war ships. A treaty was negotiated that would allow American ships access to Shimoda and Hakodate where Americans could travel within an 18 mile radius of the ports and also obtain provisions. They also agreed to allow an American consulate to be established in Shimoda. Most of the concessions that were demanded by America including the end of the seclusion policy were granted. The only thing the treaty didn't grant was trade. In 1856 the Europeans demanded and received similar treaties to that granted to America.

The events of 1856 produced much controversy in Japan further destabilizing the Shogunate. In 1858 the Chinese once again suffered a crushing military defeat by the western powers. As a result of their defeat they agreed to a new treaty embodying further concessions with Britain, which of course were automatically conferred to the rest of the members of the treaty port system. Townsend Harris who was the American Counsel General in Japan quickly made the new Chinese defeat known, and threatened that the combined western fleets were ready to set sail for Japan if a trade agreement could not be arrived at. The threat seemed credible, so Japan signed the United States – Japan

Treaty of Amity and Commerce on 29 July 1858, it provided for "formal exchange of diplomatic representatives, scheduled the opening of Edo, Kanagawa, Osaka, Hyogo, Nagasaki, and Niigata as ports and cities were foreign merchants could establish residences and enjoy the privilege of extraterritoriality. In addition, the pact placed Japanese tariffs under international control and pegged import duties at levels that benefited foreign traders" (McClain, p.142).

The treaty was detested by the Japanese and resulted in the formation of a number of aggressive factions comprised of dissident samurai and some of the Daimyo who resisted the policy. The principal Daimyo were Choshu and later Satsuma who provided much support for their fellow dissenters. The large influx of foreign trade enabled by the treaty concessions coupled with poor monetary policy caused hyper inflation, extraordinary price increases, and a 30% increase in interest rates. The hardships that the Japanese people had to endure produced wide spread dissatisfaction among the population with the Shogunate. This resulted in civil unrest characterized by rioting, plots, and assassinations of officials which became commonplace. Many of these factions wished to restore political hegemony to the moribund emperors that had been snoozing in the background for the last several hundred years. The Shogunate responded by augmenting its military power to shore up its position through repression. In 1867 during an unsuccessful military campaign against dissenting Daimyo, the Shogun Iemochi died. The new Shogun continued with a hard line militaristic program and was defeated by the Daimyo coalition. In order to gain legitimacy, the Daimyo coalition needed to acquire the support of the 15 year old Heavenly Sovereign. On 3 January 1868 Satsuma warriors captured the royal compound, and were met by antishogunal courtiers, later that day the Heavenly Sovereign issued a proclamation abolishing the shogunate. Edo was captured by the new imperial army in April 1868 and the last shogunate forces were mopped up in 1869.

The members of the new government came to the conclusion that the only way to achieve equality in the western dominated international community was to establish economic and military parity, as well as a political system modeled on western governments. They set out to accomplish this task by subsidizing industrial development and modernizing their military which adopted many features of the Prussian system. However, Japan is deficient in some of the resources necessary for this type of economy and had to acquire them from offshore sources, thus the era of a sustainable society came to an end. During this same period named the Meiji, the Russians were maneuvering to acquire some of the Japanese northern island territories. The Japanese response was to conquer Korea to establish a buffer zone. The Koreans were quickly exploited in a way similar to that which Japan was being exploited by the western powers. That is, the Koreans were forced to grant treaty ports with the privilege of extraterritoriality. In 1879 the Japanese decided to "straighten out their boarders" and annexed the Ryuku Islands which became the Okinawa Prefecture. In 1895 the Japanese started to make incursions into China because they needed a new "buffer zone" to protect the Korean buffer zone from Russia and of course obtain more resources (Treaty of Shimonoseki with China). This disturbed the western powers who forced the Japanese to relinquish most of their gains in China. In 1894 the existing commercial treaty was renegotiated with Britain, who needed Japanese good will to contain Russian ambitions in Asia. The new treaty abolished extraterritoriality within five years. In 1897 the other western powers recognized that the Japanese had considerable military strength, as a result the other treaty powers agreed to similar terms and also recognized Japanese tariff autonomy. By 1911 normalized relations were established between the western powers and Japan. Normalization resulted in the detested treaty provisions imposed on them by Perry to be completely abolished. In 1904 tensions between the Russians and Japan resulted in Russo-Japanese war. The Russians experienced a crushing defeat (Treaty of Portsmouth 1905) which forced Russia to relinquish territory and commercial resources. Later in the year 1905 Korea became a Japanese Protectorate. The trajectory of imperial expansion continued being justified by the ideology of creating an "Asian co-prosperity sphere" which eventually embraced large portions of China and other pacific basin nations. This expansion alarmed the western nations who sought to curb Japanese power. This was accomplished by the imposition of military and trade restrictions, which eventually resulted in World War II with the Allied Powers.

If we consider these societies a common theme can be discerned. During the Meiji and the following early modern period, the Japanese needed to emulate western practices to regain and retain their autonomy. To carry out the westernization program required the extraction of resources from external areas since their own resource base was inadequate. Once the growth based laissez fair capitalist economic system was implemented a positive feed back loop of consumption is created. Its internal logic always requires acquisition of additional resources to sustain its expansion. These resources come from newly acquired areas, and are used to further feed and grow the industrial expansion, which must be protected requiring further resources, thereby, fueling further expansion …....
continuing in an increasing spiral of acquisitiveness and

extraction, which invariably requires the use of violence to implement. The increasing levels of violence produce a need for additional military resources. To provide these munitions requires expansion of their armaments industries, which then need increasing supplies to support further military expansion in order to consolidate their grip on the new territories to extract more and so on. Isn't this essentially the pattern described by Bhikku Bodhi?

If we consider the first three societies used as examples, they all were unsustainable because the consumption of available resources exceeded what was available. In addition to engaging in excessive resource consumption, these societies used agricultural practices that depleted soil fertility which reduced harvest sizes. These maladaptive activities were further exacerbated by population expansion which placed increasingly greater demands on their resource base. The Easter Islanders who were isolated on their island obviously did not implement corrective measures to adjust their level of consumption to what was locally available. The inland Maya kingdoms suffered the same fate as the Easter Islanders. Their isolation while not complete but was greatly reduced by endemic warfare that limited travel out of their kingdoms which occupied small areas. The Maya also restricted the use of the wheel to religious activities, which produced further limitations on transportation. A few of the Mayan kingdoms for example, Lamanai and the island traders who enjoyed diverse resource bases provided by more effective forms of transportation, were less sensitive to environmental constraints and were able to continue into post classic times. In Colonial America the colonists had effective forms of transportation, but were bumping up against the same problem of exceeding their resource base. The resource base became inadequate as a result of soil degradation that produced falling agricultural productivity which was exasperated by the increasing demands of an expanding population. Relief of these structural problems in colonial American society were not addressed and solved even though methods of maintaining soil fertility were known.[33] Thus, the colonists where in a bind produced by constraints in expansion, which was their only remaining avenue of relief. The restriction of colonial expansion originated from English trade policies (mercantilism) and treaty agreements limiting the territorial area. The American colonists were able to eventually relieve the limitations of their resource base by engaging in a number of wars of conquest. Eventually the continental wars of westward expansion on the American continent came to an end at the Pacific Ocean. Once the expansion was checked by the Pacific Ocean, a new source of materials needed to be acquired. Thus, attention was shifted offshore and the theater of military activity was expanded to include overseas acquisitions. A few examples of this imperial activity was the conquests of Hawaii, Panama, and the larger Spanish American war which expanded American territory to include the Philippines, Puerto Rico, Guam and other smaller acquisitions. More recently a much greater level of exploitation of external resources has engendered the use of the military to acquire control of

---

[33]   *The European nations that the colonists emigrated from had developed methods to help sustain soil fertility. Many documents exist describing methods of crop rotation, appropriate levels of livestock loading, etc. At Mt. Vernon George Washington had a masonry pit constructed for aging manure which is still in existence. Many articles and pamphlets were printed describing methods to retain soil fertility in colonial America, demonstrating that this knowledge existed. Unfortunately these practices were rarely used by the colonists most of whom adopted the practice of disposability.*

Iraq, Afghanistan, and Panama. Because of public opposition to the application of military force for the benefit of narrow interests, the use of overt aggression has been supplemented by less blatant uses of force to mask the activity of acquisition. To reduce public awareness of these activities lower profile methods that employ proxies, fermenting revolutions, or by creating client states are used. For example, the Contras, Iran in 1953, and Saudi Arabia respectively.

Prior to the Edo period the Japanese where following the same pattern of expansion and over exploitation as described in the first three case studies. They had severely deforested their country resulting in the familiar pattern engendered by the excessive use of resources. However, officials of the Shogunate recognized the developing imbalance between resource usage and availability and took action. They formulated measures to establish a sustainable resource usage pattern, as well as to reforest the nation. Their programs were successful, producing a one to one balance between usage and consumption for 265 years. The United States disrupted the sustainable society established by the Shogunate by compelling the Japanese to participate in the exploitative form of economic system organic to the western powers.

In general the elites of our case studies formulated and implemented ideologies that supported their positions. Easter Island was controlled by a small number of chiefs who had hundreds of gigantic statues erected to produce a tangible reminder of their legitimacy. Legitimacy was an essential enabler to support the projection of ancestral power to exert control of the wider population. The Mayan kings legitimacy was based upon an ideology that portrayed them as the conduits to their various gods. Their influence with the gods was a result of their blood relationship to their ancestors who had undergone a metamorphosis after

85

their deaths into said gods. These gods controlled natural processes that provided the means of survival, thus supporting the royals was essential to life. Like the Easter Islanders, the Maya elites diverted large amounts of resources to produce monumental architecture for personal and ancestral aggrandizement as a display of power. In the United States the elites formulated and promulgated the ideologies of racism and manifest destiny to provide a basis for an aggressive policy of exploitation. In current American society, ideologies of being a conservative or liberal are being implemented through sophisticated PR campaigns to maintain the status quo, by directing public attention and energies away from looming problems. This is being accomplished by use of what in essence is a strategy of divide and conquer by producing polarization. The Japanese ideologies were based upon the semi-divine nature of the emperor whose lineage originated from the sun goddess Amaterasu Omikami. The "blessing" of the emperor gave legitimacy to the activities of the elites, the shogun acted on his behalf and in the post shogunate era policy was dominated by various powerful cliques who were "implementing the imperial prerogative" by formulating ideological justifications for expansion such as that articulated by the Asian Co-prosperity Sphere.

Chapter 3
Humpty Dumpty

Humpty Dumpty sat on a wall,
Humpty Dumpty had a great fall,
all the kings horses and all the kings men
couldn't put Humpty together again.
Children s nursery rhyme.

Presently we are being faced with a number of global scale problems. Rapid change is taking place in the global climate, at the same time many critical resources are or becoming depleted. A rapid expansion of the worlds population and an increasing level of resource usage is also occurring, providing the underlying driving forces for many of these problems.

Climate change has and continues to be a controversial topic as a result of uncertainties. These uncertainties are produced because of its extremely high level of complexity which gives rise to variation in expert opinion about the details of projected changes. If you examine the variations of anticipated climate change, the various scenarios are all fairly similar except that they occur over different time periods. The range from the most pessimistic climate projections to the optimistic is about thirty years. If one surveys the climatologists literature the only significant amount of controversy is about when these changes will take place not what they will be. Contrary to the PR being parroted by the mass media the number of dissenters in the scientific community doing industry independent research on this topic is practically nonexistent and has been for quite some time.

The facts that I found convincing about climate change were related to the oceans. After the Soviet Union collapsed the United States government released information that the navy had collected about North Polar ice cap thicknesses. This information was crucial to the navy who had a fleet of specially designed submarines that could break through the polar ice in order to launch a missile. These submarines had a limitation though, they could only break through the thinner ice. Thus, for these ships to be able to break through the ice and launch a missile the ice thicknesses had to be known. To provide this information continuously updated precision maps were produced of ice thicknesses during this period (from the early 1950's - 1990's). The maps showed that during this period the average thickness of the ice diminished by about 40% which can be correlated to changes in temperature producing greater annual melting.

Further evidence can be found if we look to the history of polar exploration. Almost from the very start of the arrival of Europeans in the new world they launched expedition after expedition to try to find a north-west passage for ships to sail above Canada. If such a route could be found, it would have provided a much shorter distance than going around South America to trade locations in the far east. They all failed until the 1903 Norwegian expedition.[34] because of the extent of polar ice on the northern Canadian coast. The Norwegian expedition succeeded because it utilized a small shallow draft ship that could take advantage of small patches of open water that occur near the shore during summer. The expedition required three years to accomplish! Now the north-west passage is being used

---

[34]   *Almundsen expedition from 1903 – 1906.*

occasionally for commercial shipping because of the shrinkage of the ice cap due to melting.

Sea level rise as a result of thermal expansion combined with contributions of water from glacial melt, is another direct indicator of global warming. The height of the sea has been recorded for centuries from tidal gauge data, engineering information used to construct harbors, locks, levies, dykes and other civil engineering projects. More recently satellite altimetry has provided very accurate measurements of nearly the complete globe (Fig. 1). These measurements indicate that sea level was stabile until the late 19th century. Starting in 1870 to now it has risen approximately 21.6 cm (8 ½ in.). From 1961 – 2003 the average amount of sea level rise has been 1.11 mm/yr, (.42 mm from thermal expansion and .69 mm from glacial melt.[xxii] The rate of sea level rise has been increasing, if the period from 1993 – 2003 is extracted from the above average the amount of rise from thermal expansion and melt has increased to 1.6 mm/year (1/16 inch/year).

Thermal expansion as a result of temperature increase is an easily and directly measurable property of all types of matter.[35] For example, if one looks underneath bridges the girders are supported by rollers to allow for movement produced by thermal expansion, in industry railroad wheels are heated up to cause expansion and then allowed to shrink by cooling onto the axle providing a very strong mounting.

---

[35]   *With the exception of when water freezes which is caused by the arrangement of the water molecules into a crystal lattice which has an expanded arrangement of the molecules.*

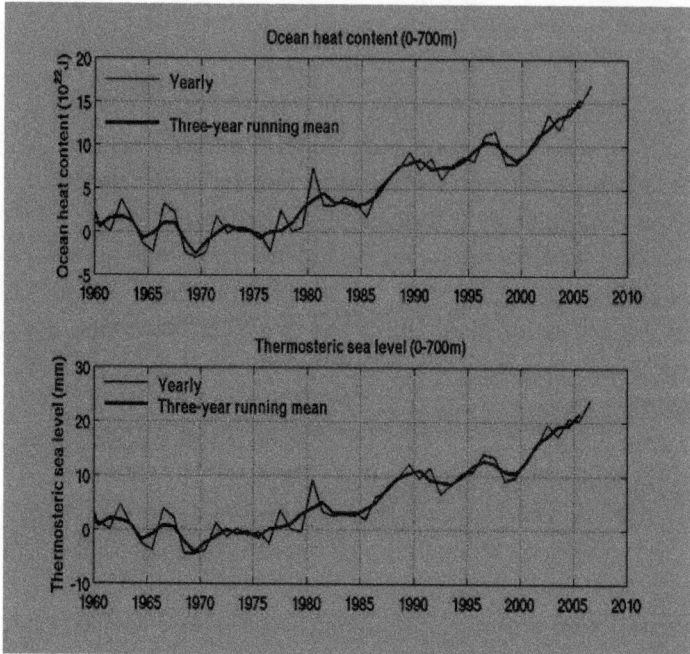

Fig. 1 Source SOTC

If we consider these first three examples they are all the result of direct measurement or written historical records based upon direct observation provides a solid factual support indicating that global warming is indeed taking place.

The next step is to link the warming trend noted above to "green house gases" which act to retain heat in the atmosphere. The atmospheric temperature increase is a result of these gases ability to reduce the escape of heat produced by the sun shining on the earth into outer space. There are a number of green house gasses that are produced by human activities, the most significant of these are carbon dioxide, methane, water vapor, nitrous oxide, and chlorinated floural carbons (CFC's).

Carbon dioxide is the most significant because not only does it produce the green house warming effect but it's warming activity is also a contributing factor in the increase in methane and water vapor. If the concentration of atmospheric carbon dioxide is considered it remained at 265 – 280 parts/ million (ppm) until the late 19th century.[36] Recent direct carbon dioxide measurements have been taken at Mauna Loa Hawaii which is an observatory located on the upper reaches of Mauna Loa volcano. The observatory was located there because the air composition is not influenced by local contaminants and can be used to represent the base composition of the entire atmosphere. Assays of air samples have been conducted at this location since 1960. As a result of these measurements we know that the current amount of carbon dioxide in the atmosphere is around 398 ppm (.0398%) an increase of about 37% over the 288 ppm (.0288%) value or less that has lasted for the last 10,000 years (Fig. 2).

---

[36]   *The gas composition of the past are derived from ice cores obtained from Green land and the Antarctic. Each year new snow falls and traps air in openings between the snow flakes, during the summer the upper surface melts, thereby, producing a yearly series of easily discernible layers. The trapped air gasses can be collected from any past year simply by counting back the number of layers to the desired date. It's a simple matter to remove the air gasses from the ice sample all you need to do is warm it up (you can do this yourself by taking a sauce pan and heating up some water, you will find bubbles of dissolved air forming in the water before it boils). The air bubbles are then collected and chemically analyzed. The temperature of the water that evaporated from the surrounding ocean to form the clouds that produced the snow can also be determined by analysis of the amount of an isotope of oxygen in the water (oxygen 18) which varies with water temperature.*

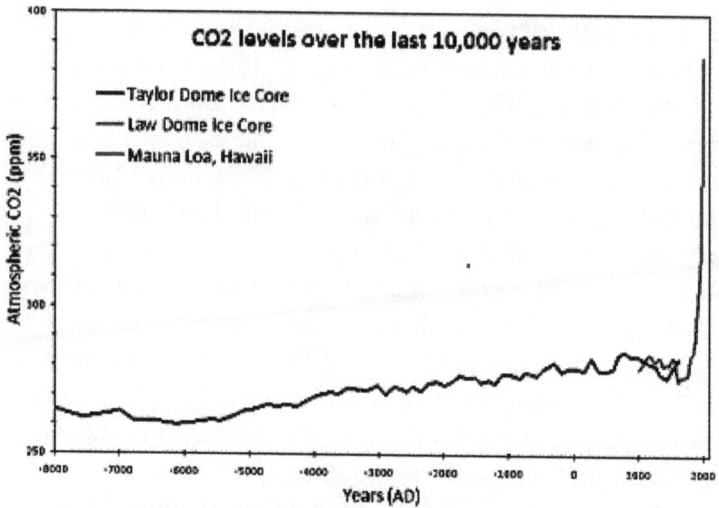

Fig. 2 Courtesy of Skepticalscience.com

How do we know that carbon dioxide increases air temperature? This can be established very simply, all that needs to be done is increase the amount of carbon dioxide in the air in an enclosure. You simply place the enclosure outside and expose it to the mid-day sun and measure it's temperature with a thermometer and compare it to the surrounding air. If you wish to verify this here's a simple way to show this effect (see footnote for test procedure).[37] Now that we have proven that carbon dioxide induced atmospheric heating is a fact using simple direct

---

[37]   *If you wish to verify this for yourself, here is an experimental setup: obtain 2 small inexpensive coolers and place the same amount of some dark material inside each one to absorb the sun, in one mix some baking soda and vinegar together which produces carbon dioxide and quickly seal the top with clear plastic wrap, take the second cooler and seal top with plastic wrap (no vinegar & baking soda) place outside in the sun for awhile and measure the temperature in each one. You will find that the one with more carbon dioxide has a higher temperature .*

measurements we can consider Fig. 3. Figure 3 provides a comparison of the amount of carbon dioxide increase to the world's temperature increase, i.e., global warming.

As we can see in Fig. 3 average global temperature increase closely follows the amount of carbon dioxide increase in the air. Moreover, the temperature increase has also been verified by sea level rise, which corresponds to water expansion plus increases due to melt water. Furthermore, the increases in sea level from melt water matches the reductions in the amount of glacial and polar ice. The last step is to demonstrate that human activities are producing the increase in atmospheric carbon dioxide.

If we consider Fig. 4 we can see that there is a further close correlation between the amount of carbon dioxide being produced primarily through combustion of fossil fuels which are released directly into the air and the amount of atmospheric carbon dioxide increase.

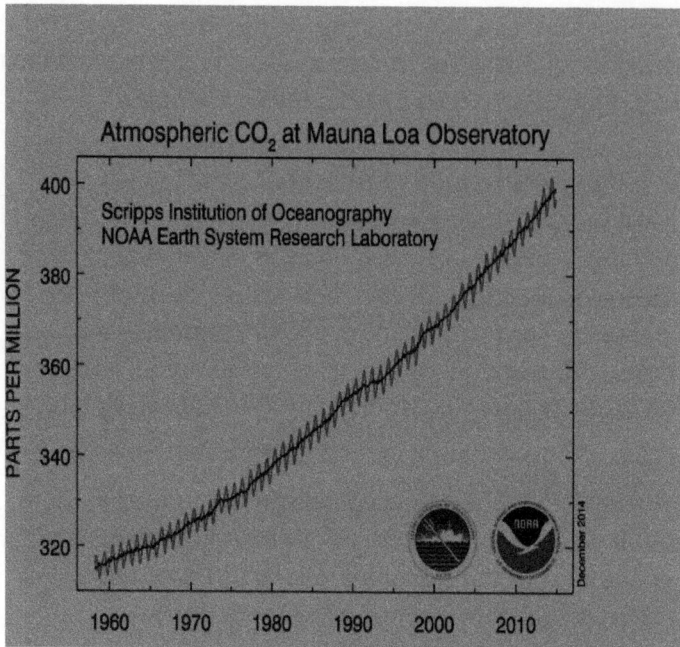

Fig. 4 Courtesy of NOAA.

Most of the carbon dioxide that we humans have been adding to the air is absorbed by the oceans, and a trivial amount by other bodies of water. So far this has prevented a much larger global temperature increase which is good. The absorption of carbon dioxide by water has mixed blessings though, some of it is changed into carbonic acid. Carbonic acid is a weak acid in the same strength range as the acid found in vinegar or lemon juice. The strength of acids like carbonic and bases such as dilute lye solutions are measured using pH, which has a range of 1 to 14, seven is neutral and each number represents a change of strength of 10 times. In the last 100 years the average pH of ocean water has changed from pH 8.16 to 8.07[xxiii], which indicates a 30% increase in acid strength. If you recall the baking soda and vinegar experiment in foot note 33, chemically the baking soda is named sodium bi-carbonate. The shell of shell fish is comprised of a substance whose name is calcium carbonate, you will note that the baking soda is part carbonate as well as the shell fish shell. Acids dissolve the carbonate in shell in the same manner as the baking soda. The result of the acidification of the oceans is that shell fish are having difficulty surviving and have vanished from some places. At the present time tests are being conducted to see if large amounts of crushed clam shells placed in defunct shell fish beds in Main are being conducted. This is being done in the hope that the crushed shells will scavenge enough of the acid locally from the sea to allow shell fish to be able to grow there again.[xxiv]

Methane is another of the green house gasses, 40 % occur from natural sources and 60 % as a result of human activities. It has 25 times the warming effect per molecule than a molecule of carbon dioxide.[38] In the United States it accounts for 9% of the green house gasses that are produced and remains in the air for 12.4 years. The natural pre-industrial amount of methane in the air was 715 ppm. and has risen to a present value of 1,774 ppm as of 2005.[39] The sources of methane are :

Natural Gas & Petroleum.................30%

Ruminants livestock[40]........................23%

---

[38] . *Since 1990 the United States EPA currently lists methane's green house strength as being 21 times greater than carbon dioxide, however recently the EPA announced that it was upping the value to 25X based upon more recent research.*

[39] *The natural range of atmospheric methane for the last 650,000 years varied from 320 – 790 ppm.*

[40] *The ruminant animals are: Cattle, sheep, goats, buffalo, and camels.*

Landfills..17%

Coal Mining.......................................11%

Manure...9%

Misc...7%

Waste Water Treatment.......................3%

The 40% of methane that is produced from natural sources mostly comes from decaying vegetation and other organic material by bacteria that thrive in oxygen free environments, primarily from wetlands. Some other small natural sources are termites, oceans, volcanoes, and wildfires.[xxv]

Water vapor also acts as a green house gas, the amount present in the atmosphere is dependent upon temperature. The warmer the temperature the greater the amount of evaporation and also the ability of the air to retain moisture. This phenomena can be easily seen when one takes a "tall cool one" from the refrigerator on a hot humid day. The warm humid air is chilled on the side of the cool glass and can not retain as much water vapor. The water vapor condenses out on the glass causing the formation of water droplets. Recent research[xxvi] has confirmed that water vapor traps a lot of solar energy. Dressler a researcher at Texas A&M university has analyzed the new Aqua satellite [41] data compiled from 2003 – 2008 said that, "water vapor feedback is extraordinarily strong, capable of doubling the warming due to carbon dioxide alone."

---

[41]   *The Aqua satellite has the capability of measuring the humidity throughout the air column from the ground to an altitude of 10 miles. It has allowed a great refinement of the state of water vapor profiles over ground based measurements alone.*

Nitrous oxide is also a significant greenhouse contributor, it is primarily produced by agricultural activities. 69% of it originates from the use of synthetic nitrogen fertilizers, especially of the ammonia types. An additional 5% originates from manure, as a result of anaerobic (oxygen free) fermentation. The conversion of fertilizer into nitrous oxide occurs after it is applied to the fields by soil microbes which convert some of the ammonia into this gas. This gas leaves the soil and is released into the air. The remaining amounts come from industry (9%), stationary combustion (6%), transportation (5%), and miscellaneous (6%).[xxvii] This gas has 310 times the heat trapping ability of carbon dioxide and remains in the air for 114 years. The amount has increased from a pre-industrial level of 270 ppb to 319 ppb as of 2005.[xxviii]

Another group of greenhouse gasses are chloroflourolcarbons (CFC's) that were developed for use in refrigeration. The use of these substances were later expanded as a means to pressurize spray cans, and as cleaning agents. They are definitely a significant retainer of heat in the atmosphere, and are comprised of a number of compounds some which have thousands of times of the warming potential of carbon dioxide. CFC's can persist in the atmosphere from a few decades to thousands of years. At the present time, estimates of the contributions to atmospheric heating ranges from being significant to major. The precise amount of heating is controversial though, and we will have to wait to see how things develop. These compounds are also the ones that destroy ozone which protects us from excessive ultraviolet radiation. Production of these compounds where banned in the 1980's by first world countries. Some of the third world nations are still producing these substances because they could not afford to replace their refrigeration. As a result of the CFC ban the ozone hole over the antarctic has recently started to shrink.

Predictions by experts indicate that the ozone shield should be largely restored to its original size in about 50 years.

So far we have considered directly measurable human induced inputs that can be seen to be producing the current elevation of global temperature. Next we will try to peer into the future by examining what the climate modelers are predicting. Climatologists have developed sophisticated computer models for climate prediction. The way these models are tested is by starting them using the current weather conditions and then running them backwards. These models are perfected by "tweaking" them until the computer model can accurately produce predictions that correspond to the known past climate. Once the model is able to predict the known past climate it is then run forward to predict the future climate. The climatologists examine different scenarios by varying some of the factors the computer program uses such as the amount of carbon dioxide, methane, etc., that may eventually be present in the air. Generally a number of these models are run a number of times using the differing values and then averaged. Since the climate is extremely complex and the effects of the inputs are highly variable these models are not able to produce very precise highly detailed results.

The fundamental driving force that produces weather is caused by the temperature differential between the equatorial regions and the polar regions. In the warm equatorial regions of the earth the air is heated which causes it to rise. Once it arrives at the poles where it is cold the air cools and sinks down. In order to keep providing air to supply these vertical movements the warm air arriving at high altitudes is pulled toward the polar regions to replace the cooled air that has sunk down. At the same time the cool air from the poles move at low altitudes toward the equatorial regions caused by air pressure differentials. A simplified way of visualizing this is that these wind currents are somewhat like a conveyor belt, with the warmed air on top moving toward the polar regions, and the cooled air on the bottom going back to the equator. Recent research by Francis et al., have found that increases of air temperature caused by global warming has produced faster warming of the arctic. Faster arctic warming has caused a reduction of the temperature differential between the warm southern regions and the arctic. At the present time the arctic is warming at about twice the rate relative to the more southern portions of northern hemisphere.[xxix] Since the arctic is becoming warmer relative to the southern regions, the north south air circulation is weakening. As a result of the weaker circulation, the behavior of the jet stream which largely governs our weather is changing. The jet stream progresses from west to east and generally fluctuates up and down in waves with respect to its latitude. As a result of the lower temperature differential the jet stream circulation pattern has become more torpid, the progress of the fluctuations have diminished. The slower progression of the wave fluctuations result in weather patterns that are more persistent. Greater persistence of weather patterns has the effect of increasing the likelihood that "certain types of extreme weather, such as drought, prolonged precipitation, cold spells, and heat waves" (Francis, p 3) will take place.

The oceans are the other greatest contributor to weather by providing moisture and acting as a sink for a great deal of the pollutants we have been producing. They have been absorbing 95.5% of the excess heat that is being produced by the increasing amount of green house gasses in the air (see Fig. 1). Like the convection currents in the air forming a "conveyor system" the oceans have an analogous set of circulation patterns. The warmed surface currents move poleward and upon their arrival sink down. After sinking down they head back to the warming regions along the bottom of the ocean (Fig. 5). A familiar example of this system is the Gulf Stream which travels off the coast of North America, crosses the Atlantic Ocean, and goes past the United Kingdom. During it's travels these surface waters are evaporating which increases the salt content producing greater density. The salt content is further increased as a result of it's northward progress to colder areas where the formation of ice withdraws some additional fresh water. A further increase in the density of the salty water is produced by reductions in it's temperature. The combination of these factors eventually causes it to sink down into the abyssal depths in the vicinity of Green Land and is referred to as the thermohaline circulation. These warm surface currents have a significant impact upon the climate in their vicinity. In the United Kingdom and in other nearby coastal areas the climate is moderated, in fact in Ireland one can find palm trees growing even though it is at about the same latitude as Wisconsin where last winters temperatures were occasionally -25 F.

Recently there has been conjecture within the scientific community about the possibility of the thermohaline circulation being affected by the introduction of the large amount of fresh water originating from melting of the

Global Ocean Conveyor Circulation

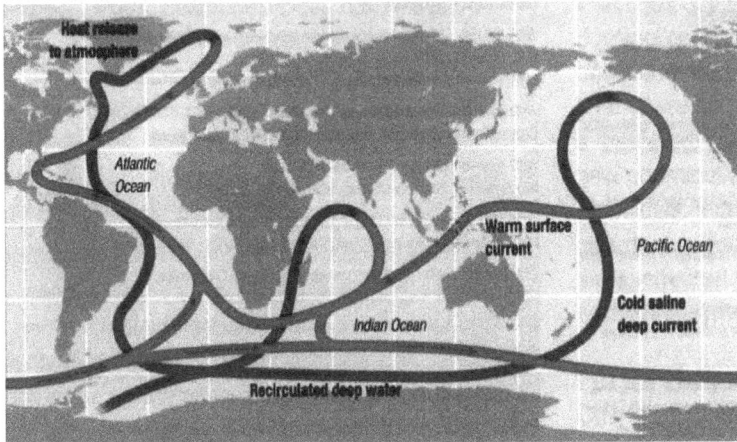

Fig. 5 Thermohaline Circulation
Reprinted with the permission of the United Nations.

north polar ice cap and nearby glaciers. The addition of these large quantities of melt water is reducing the salinity in the northern waters. If the density differential becomes too small the thermos-haline circulation could stall, producing a profound effect upon the climate. However, at the present time there is no convincing evidence that the rate of circulation has changed beyond it's normal level of variation.

In general what are the climatologists predicting? A greater variation in the climate accompanied by persistent weather regimes as described by Francis and her colleague. A greater percentage of strong storms, and in the regions where the temperature falls below freezing greater impacts from freezing and thawing cycles on infrastructure. Changes in the distribution of the amount of precipitation and normal ranges of temperature. The 2014 UN Intergovernmental Report on Climate Change[xxx] indicates that in the next several decades we can expect to see a range of change in precipitation of about 35%. Some areas will experience an increase as much as 10% in precipitation and other areas a decline of as much as 25%. The net effect on food production from this source is expected to be a 10 - 15% reduction in yields. The costs of maintenance for infrastructure will increase. In the temperate and subarctic climatic areas, diseases, pests, plants, and animals that are presently found in warmer areas will migrate to higher latitudes and elevations. The presence of these novel species are and will increasingly produce disruptive elements in the current ecological and agricultural regimes. For example, since the 1990's the mountain pine beetle that is adapted to warmer climates has been migrating north into the Canadian forests causing widespread damage in British Columbia where 50% of lodge pole pine trees have died from infestation. Recently these beetles have migrated into Alberta.[xxxi] They have also begun to eat tree species that were not favored by them before, such as jack pine in Canada. In Colorado where another beetle epidemic is underway these insects have started to eat ponderosa pine according to the US Forest Service. These beetle infestations are attributed to the warming temperatures that have allowed them to expand their range into areas where low winter temperatures prevented them from becoming established. The cold adapted northern tree species have no natural defenses to fend them off. Recently gardeners in northern areas have reported southern pest species appearing in gardens such as the tobacco bud worm.[xxxii]

The question is, how much reliance can be placed upon these mathematically modeled climate predictions? According to the 2014 UN IPCC report these models have the capability to provide "hind-casts" with some skill of the near surface temperature over much of the globe for 9 years (IPCC, p. 965). However, there are many unpredictable events that can cause these models to diverge from their predicted path. For example, large volcanic eruptions eject large amounts of particles into the upper atmosphere which reflect sunlight and cool the earth. It is my view that these models do provide some utility because there are segments of time available in forecast predictions where unpredictable events are absent or of a small magnitude (large events occur less frequently) that produce gradual divergence from it's prediction. In the final analysis their ability to hind-cast does validate the predicted effects of greenhouse gasses on climate since the effects of greenhouse gasses are integral to the functioning of these models.

Status of the Earths Agricultural Potential: At the present time about 1/10 of the earths land is desertifying, and about 1/3 of the earth is changing to dry-lands that can easily become desert if the current trends continue. According to a study done by Pimental, in the last 40 years erosion has rendered 30% of the worlds arable land unproductive for agriculture.[xxxiii] A recent study by The National Academy of Sciences has found that cropland soil is being eroded at 10 times the rate of soil formation, about 1% per year. In the United States it takes 500 years for 25 mm (1 inch) of top soil to form. These agricultural losses are further being augmented by global warming. Recent research has demonstrated that a 1 degree centigrade increase in the minimum temperature during a growing season produces a 10% reduction in yields of rice, wheat, and barley.[xxxiv] The result of these trends is that since the 1980's the amount of land under cultivation has started to diminish, and agricultural production has started to contract relative to the rate of population expansion. The population growth rate between 1990 – 2000 was 1.3% per year and food production increased by 1.2% per year[42] the first time this has occurred in human history!

In recent years much corporate PR has ballyhooed the wonders of genetically modified organisms (GMO) crop plants and animal products, the general theme is that these products will ward off world starvation.[43] Recent studies of GMO crops have shown that not much difference in yield exists when compared to conventionally grown non-genetically modified crops. However, the GMO crops have the disadvantage that they require greater inputs of costly synthetic agricultural chemicals. If one considers the quality of the studies provided by the producers of these materials they are inadequate to formulate any realistic appraisal of their safety. Basically these studies are restricted in scope to produce results that support the industries PR gloss.

The United States Government accepts these superficial studies as a result of politically engendered limitations placed upon the regulatory apparatus. The approval of these products by the regulatory agencies is then utilized by these industries in their PR campaigns to foster the impression that these products have been thoroughly vetted by the government and no need exists to further probe their safety for deployment and consumption. When

---

[42]   *Much of the 1.2% increase is a result of growing additional crop(s) for harvest in a year on the same land and not a result of greater food output by plants, thereby, masking the effects of reductions in arable soil.*

[43]   *At the present time the amount of food being produced is greater than needed to support the worlds population even though wide spread starvation is present. The problem is that distribution of existing supplies isn't functioning adequately.*

consideration is given to how government policy is formulated, the impression one gets is that the United States government is a rubber stamp for anything big business desires.

The question is, why are these studies inadequate? To formulate their PR the biotech industry has persisted in utilizing an incorrect outdated genetic model. This model uses a scenario where a gene has a simple one to one form of direct control of the synthesis of a particular biological agent, where one gene produces one protein. Recent discoveries indicate that this is not the case, the system of genetically controlled synthesis is much much more complicated. What has been found is that the instructions produced by a gene is modified by spliceosomes that rearrange the genes instructions, thereby enabling a single gene to produce hundreds or thousands of different proteins.[44] The complexity of this system is further augmented through the addition of extra molecules to the newly produced protein. These molecular additions are governed by the environment where protein synthesis is occurring and changes the basic protein into an entirely new substance. A further layer of complexity also arises as a result of another class of molecules that control the architecture (shape) of the protein which is essential for it to function. Needless to say it is extremely difficult to unravel these highly complex poorly understood processes, and therefore prohibitively expensive. Also, when one considers that genes from organisms that are not even remotely related are being combined such as those of fish and plants, it is unknown how the new synthetic organisms molecular machinery will function. In fact it has been

---

[44]   ..WE know that this is the case because many more types of proteins exist in an organism than there are genes.

found that many of these products have unanticipated novel new molecules that have unknown effects. Basically, the planetary environment and public are being used as guinea pigs by these industries. Since these products incorporate a novel type of composition they have produced new types of diseases such as eosinophilia myalgia syndrome[45] and other allergic reactions. Since these medical problems are originating from novel sources, the diagnostic protocols and any potential treatment modalities are unknown requiring development, if possible. The GMO industry has also widely introduced these new and highly novel organisms into the environment. They are working on the assumption that because they externally resemble existing species that they will function in the environment the same way, in spite of the fact that they are specifically being produced to be biochemically and functionally different. For example, consider the difference between salt and sugar. Both have the same appearance but are quite different. If one looks into the status of the understanding of soil ecology, what is actually known can be characterized as being at best rudimentary. These businesses are producing and promoting products that incorporate many unknowns with the potential to produce environmental damage. This type of damage can take a significant length of time to become recognizable and may not be correctable because of the complex interdependent nature of ecological systems.[46] I wonder, do you think the

---

[45]   ..Apparently this disease was produced as a result of the additional production of unanticipated molecular by-products in a GM process used to produce L-tryptophan produced by Showa Denko as a supplement (this has been removed from the market).

[46] In fact we are already starting to see signs that this type of environmental damage is occurring. GMO plants

people who are getting rich from these enterprises will step forward and expend their own resources to remedy any problems or will the public have to pay?

And in the final analysis we don't really need GMO at the present time since enough food is currently being produced. To address the impending food shortfalls in the future other means exist that can be used to solve this problem (this will be discussed later).[47] In my opinion a prudent course would be to restrict the implementation of this type of technology to clinical or other thoroughly controlled settings.[48]

---

that produce their own insect poisons (referred to as BT) have been shown to produce greater negative impacts on beneficial insects such as honey bees, lacewings, etc., than on pests.

[47] *With respect to carrying out this type of research, I think it holds great promise but should be limited to just research which could be tested in something like the biosphere and not released into the environment. Our understanding is simply too limited at the present time to implement it. Therapeutic applications also have a great beneficial potential and should be investigated vigorously and implemented after through clinical trials. This type of application of biotechnology does not carry the same types of risks as GMO crops but needs improved controls to prevent problems like EMS which was mentioned above. It may be a better approach to limit GMO research to government sponsored efforts instead of businesses whose sole purpose of existence is to make quick profits for their owners and where ethics at best is a tertiary consideration.*

[48] *At the present time citizen ballot initiatives are being worked on in Oregon and Colorado for GMO labeling.*

Let's consider a few anecdotal GMO case studies. In my own case I started having symptoms of excess mucus production in the nasal and throat areas as well as a build up of pressure in the ears. The build of ear pressure at lower levels impaired my hearing. A sever attack would cause internal ear pain, nausea and loss of balance where the surrounding area was "spinning around." During the more sever type of attacks it was difficult to move around, being reduced to a style of crawling that relied on visual clues (looking at the position of my arms & legs) to know the position of my body. I noticed that these occurrences usually took place after a meal which aroused a suspicion that there may be a connection to the food. To test this idea I systematically eliminated the various types of food I normally ate and found it was due to wheat. Some time after this discovery I was chatting with an acquaintance "G" who had suddenly developed allergy problems caused by wheat too. She informed me that she had found out that GMO wheat was now being incorporated in our food and causing allergic reactions. Another acquaintance "P" had been suffering from large severe skin rashes (they resembled the types of wounds children get when they skin their knees). He mentioned that he had many medical examinations and that nobody could identify the cause of his problem. I related to him the wheat problem and suggested he may try eliminating things from his diet as I did. He did eliminate wheat and his rashes went away. Now he avoids wheat like the plague! In my own case I happen to really like pizza and started making them with a masa harina corn flour crust which worked but produced an inferior pizza. Recently summoning some courage I tried making a pizza crust from non GMO whole wheat flour and did not have an allergic reaction to it Ah, back to tasty pizza! After the pizza crust breakthrough I verified it was GMO wheat by having some soup made by a friend and ate a few commercial wheat noodles (3 or 4), the milder form of ear problem (temporary hearing impairment) returned. In Dr. Davis's recent book Wheat Belly, he argues that the huge increase in the percentage of over weight people is a result of changes in the structure of gluten that took place as a result of genetic manipulation of wheat plants. In essence the novel types of gluten present in the GMO and some hybrid wheat plants have a structure that the body doesn't know what to do with so it is makes it into fat. It also produces cravings which produce overeating usually of more wheat products which of course produce further cravings, etc.[xxxv] It would be real nice if products with GMO content were labeled so that one could at least make informed decisions about what one is buying.[49]

Deforestation: As we have already seen in the previous chapter deforestation is one of the key factors in the impoverishment and decline of societies. At the present time the same type of deforestation is taking place but on a colossal global scale. A large amount of these forest products are being used to provide materials to support the production of disposable types of products and packaging that underpins the current large scale global economic system. At the present time approximately ¾ of the original world forest resources have been cut, and in the United states 95% of this type of forest is gone.[xxxvi] According to the EPA it is estimated that approximately 45% of waste paper per year is recovered by recycling. If this percentage were increased, a reduction of one million acres of cut forest would occur as a result of every 10%

---

[49]   *When replanting logged areas with trees, the tree species are generally comprised of a limited number of species and varieties that have the greatest commercial value. In essence most of these areas are monocrops with little biological diversity.*

increase in the amount of paper recycled. What we hear from the industry is that these denuded areas can be replanted with tree farms to provide future needs. Unfortunately the amount of deforested area being replanted with trees are far less than the amount being harvested, the net rate of forest loss was 5.2 million hectares (12.8 million acres) / year from 2000 – 2010, a significant improvement over 1990 – 2000 when it was 8.3 million hectares (20.5 million acres)/ year but still a huge amount.[xxxvii] The other problem is that after three successive tree crops,[50] the fertility of the land is diminished and trees will no longer grow in these areas. In the United States reforestation has been a success. Since 1990 – 2010 an additional 7.7 million hectares (19 million acres) has been established. However, much of the older tree cropped areas in the United States are now second generation, so we can only expect one more crop of trees from these areas. Perhaps it may be a good idea to replant some of these areas with a diverse type of forest that would mimic natural forests and let this land "be fallow" in this state for a while to restore the soil before it becomes useless. Of course the time horizons for this kind of program would be in the multiple hundreds of years. What is the alternative though, to end up like the Middle East which was once heavily forested and now largely desert or follow the Easter Islanders?

Fisheries: The exhaustion of the various oceanic fisheries have been well documented and tracked. In 1943 Michael Graham of Lowestoft laboratory an expert in the north sea cod fishery published his book"The Fish Gate" which

---

[50]   *It should be noted though that imported forest products are commonly used in the US, thereby, externalizing a significant amount of US consumption.*

indicated that fish stocks were being over exploited. He formulated a "law" from experience which he named the "Great Law of Fishing". The Great Law of Fishing states that as long as the level of fisheries exploitation are unrestricted they will become commercially exhausted. Thus, to maintain the productivity of a fishery exploitation must be limited.[xxxviii] Was Graham listened to? NO! At the present time only 2 of the major fisheries in the world are not completely exhausted as a result of over exploitation. Over exploitation is a consequence of destructive fishing technologies, poor understanding of marine ecology, and misguided politically motivated regulatory choices.

To illustrate the factors producing fishery declines the following case study of the North Atlantic Cod fishery will be presented. Rosenberg et al., produced a recent study on the historical decline of the cod fishery. This study used records of landings from 1855 to 2005. In 1855 Rosenberg estimated that 1.39 million tonnes of cod were living on the Scotia Bank.

After 150 years of fishing the amount of cod had been reduced to an estimated 55,000 tonnes a 96% decrease.[xxxix] Most of us have an idealized impression of fishing as it was conducted in the past when rugged fishermen in their sou'wester went out in their dories or schooners and jigged with hand lines. Or the shell fish fishermen who went out in the bays to tong for clams or capture some lobsters from small boats. Unfortunately this form of fishing has been overtaken by industrial progress which employes evermore powerful methods of fish capture and location. Not only are the new methods much more effective at finding and catching fish but they also have become much more destructive to fish habitat and wasteful.

The advent of steam powered vessels, and later by large diesel motor ships with their great horse power have the

111

ability to tow gigantic nets with mouth openings greater than 39,000 square meters (43,000 square yards) for mid water trawling. For bottom dwelling fish dragger trawl nets are used. These nets have large heavy rollers outfitted with "tickler chains" that beat against the sea bottom as it rolls along scaring up the fish into the net opening. Not only do these rollers and tickler chains scare fish but at the same time they also destroy the underwater habitat that is needed to support future generations of fish. You are probably wondering about hook and line fishing? It's still being practiced but on a gigantic scale. For example, the tuna fishermen use what are called long lines that are 130 km (80 mi) long with thousands of hooks! These powerful fish harvesting technologies are supported by matching location technologies. Highly effective fish finder sonars that are capable of identifying fish from a single large individual all the way up to huge schools are being used. These devices are also capable of providing a precise location of the fish by using global positioning that has an accuracy within a few meters. The nets also have positioning sensors that allow their deployment at exactly the right location established by the fish finder sonar and global positioning system.

Of course as the fish and shell fish populations have declined it requires greater effort to capture fish. Until recently the response to declining fish populations was for governments to provide subsidies to enable the fishing industry to upgrade their equipment to become more effective. Recently though, it is starting to be recognized that the available fish catching capability far exceeds the capacity of the fish to reproduce. Many of the governments have started to discuss this problem at the international level and also address it at the national level where action can be taken in a more timely fashion. Unfortunately progress in reducing fish capturing capability has lagged far behind what should have been done. The underlying problem is that the fishing industry and the communities where they are based suffer economic hardships as a result of cut backs in fishing effort that results in lower amounts of landings. The fishing communities and industry are feeling the pinch, thus they invest significant resources in lobbying and providing political support to politicians to promote their interests. In many cases the politicians simply produce regulations that appear to do something but in reality provides loopholes to continue with business as usual. For example, they will allow older obsolete equipment to have levels of "maintenance" performed that basically produces upgrades to a state of the art status instead of being retired. Another dodge is to simply provide no funding for enforcement of existing regulations and/or give it a PR gloss. For example, the NOAA catch limits for 2013-2014 based upon a "fisheries disaster" that occurred in 2012 "reduces quotas for ground fish" that is cod, haddock, and flounder. The actual quotas for over half of these stocks are greater than the actual amount of fish landed in 2012 and in the case of winter flounder it was increased 150%. Furthermore, the authorities are also allowing uncaught quota from last year to be carried forward and added to this years quota. The legal size of fish that can be kept has also been reduced as well in addition to the "requirements for reporting, monitoring, and on small hand gear operations." The problem with allowing reductions in the size of keepers is that small fish have less reproductive capability than the older adult fish. The effect of lowering the size limit is to impair the future growth of the fishing stock (this is a desperation measure and very unwise). They did continue to maintain the closure of fishing areas where the fish spawn in the western Gulf of Main and Cashes Ledge which is a very good practice, although the amount of area devoted to the closed fishing grounds are inadequate. The amount of area where fishing should be prohibited by designating them as marine sanctuaries should be a minimum of 1/3 up to around 1/2 of the ocean area, according to some experts. Returning to the 2013-2014 NOAA quotas, basically for cod and many of the other species they are still being knowingly over fished to relieve the economic pressure on the fishing communities. Once again the authorities have sacrificed the greater long term good for a much smaller short term gain.[51]

One of the deficiencies in the process used for evaluation of the amount of fish and shell fish that can be sustainably caught is that the quotas have been set by measuring the amount of a specific species that are caught while ignoring the ecological system that is essential for providing a particular species needs.[52] Some other big problems are that at least 1/3 of the fish catch is discarded being comprised of untargeted fish species and juveniles (referred to as by-catch).[xl] According to (Pauly, et al.) much of the catch goes unreported, for the 6 major commercial species (Red Fish, Mackerel, Haddock, Flounder, and Hake) under reporting for all of these exceeded 20% and for most of

---

[51]   . *Even though the measures that are being taken in the United States are far from being adequate, the United States at least is making a few feeble attempts at fisheries management. Many other nations such as those that comprise the European Union are where the US was 20 or 30 years ago. A few nations such as Cuba and Iceland have done a superior job in managing their fisheries.*

[52]   *..Recently researchers have developed models that are capable of providing the type of broad assessment of much of the North Atlantic ocean basin, hopefully, this type of approach will be used to restore the marine ecosystem.*

them by 50%.[53] Typically the mostly dead by-catch is dumped overboard into the ocean not only is it a tremendous waste but it also deprives the remaining uncaught fish of a significant source of food. Because of the lack of selectivity of commercial fishing techniques the by-catch often contains threatened or endangered species. For example, the long line fishing method frequently attracts and catches sea birds such as the endangered albatross who dive down to eat, get hooked and dragged under. Dolphins and endangered turtles which need to surface to breath air drown and are also common part of the by-catch.[54] Of course the result of under reporting catches and discarding by-catch is to introduce incorrect information for use in setting the levels for fish catch quotas and the formulation of other management strategies. Another big problem is the great influence these large industries have on governments. For example, if the length of governmental time horizons are considered, they have contracted to the sort of short term planning that characterize the unsustainable extractive practices commonly employed by businesses. This kind of outlook in the regulatory community is not appropriate for the sustainable management of marine or other resources.

---

[53]   *Some nations such as Norway employ retired sea captains as observers who are placed on fishing vessels working in their waters to insure compliance.*

[54]   *. There has been some reduction in this type of by-catch in the United States TAD's (turtle exclusion devices) are required as well as similar devices for dolphins. However, not all fishing vessels are outfitted with this type of equipment since it is dependent on individual nations requirements and also the willingness of a nation to provide effective regulatory efforts.*

Concentrated animal feeding operations (CAFO's) are industrial animal, poultry, and fish production facilities (I shan't call them farms since their methods of operation bear only a faint resemblance to what is generally considered the activities carried out on a traditional farm). In these operations vast numbers of live stock are packed into the smallest possible area. It is not uncommon for the animals not to have room to lie down to rest or carry out activities that are normal types of behavior. Since these conditions severely stress the animals they often display psychotic behavior, for example, pigs bite the tails off other pigs confined in their area, poultry peck nearby birds to death. The vast quantities of manure that are produced in these facilities are far beyond the capacity for the environment to absorb. To deal with the huge quantities of manure retention ponds are created that produce severe odor, fly, and other environmental problems such as water contamination and the production of the green house gas methane. There has been instances where these ponds have failed causing the surrounding areas to be inundated with manure, not only producing huge problems for people living near to these facilities, but also draining into rivers and streams resulting in catastrophic collapses of aquatic ecosystems. In some locations these types of spills have been almost routine during heavy rainfall events that cause ponds that are filled too near their tops to overflow. Of course the people who are unlucky enough to have one of these facilities built near them find that their property values plummet, and in many cases are extremely difficult or can not be sold at all.

For example, in Wisconsin where I live several counties in the northeastern part of the state, Kewaunee and Bayfield have around 30 CAFO's. In Kewaunee county 30% of the wells have been contaminated, with some areas being as high as 50% (according to the USGS). Bayfield is located on Lake Michigan and also has CAFO problems. The part of this gigantic lake adjacent to Bayfield is now contaminated.[55] [xli] Another emerging problem originates from the fact that CAFO's routinely give sub-therapeutic doses of drugs to their animals. This practice is being used as a prophylactic to keep the animals healthy enough to go to market (the CAFO environment is an ideal breeding ground for diseases). The biologists and medical community have been warning that the use of routine sub-therapeutic doses of drugs in the CAFO environment will eventually produce drug resistant strains of microbes. Some of these microbes such as giardia, cryptosporidium, and legionella are types of water born diseases that have been found in CAFO's and are challenging to eradicate since they can not be killed by chlorination.[56] This is important because chlorination is the standard method used for removal of biological contaminates in water.[57] If these

---

[55] *In 1993 The city of Milwaukee Wisconsin had an outbreak of cryptosporidium that is believed to have originated from contaminated water produced as a result of runoff during a heavy rainstorm. An estimated 403,000 people became ill, 4,400 were hospitalized, and 69 people died.*

[56] *Recently I heard anecdotal evidence that a new resistant strain of giardia has appeared in Kewaunee county from a resident. So far I haven't found any information on this in the peer reviewed literature.*

[57] *Several methods do exist to remove these organisms during water treatment in municipal systems. Final treatment with ultra violet radiation, ozone, and high level sedimentation (sedimentation is not 100%*

types of organisms become established in the aquifers, the water problems in these areas will become greatly exasperated (contamination of a well is usually corrected by chlorination). If you wish to delve into this further, greencdf.org a group I am working with has an excellent science paper available on this topic.

If oceanic fish farms that grow carnivorous fish such as salmon[58] and shrimp are considered they have many of the same problems caused by over crowding as their land based cousins. Generally the fish pens are located in estuaries that are important spawning and growth areas for wild juvenile fish. They are also frequently established in prime habitat locations where the large fish like to congregate. Further problems arise from the highly concentrated captive fish populations in these areas which become heavily contaminated by feces and rotting fish food that falls to the bottom polluting the environment and depriving the wild fish populations use of these spaces. The farmed fish have also become incubators of diseases and pests such as sea lice which attach themselves to the fish and gradually eat them. Again in order to control the frequently occurring diseases in these facilities large quantities of antibiotics are used in the same way as in land based CAFO's. As a result of the frequent widespread usage of antibiotics in these operations, experts in the medical community are expressing concerns about the likelihood of

---

effective). All of these methods add to the cost of water treatment.

[58]   *Salmon are the most prevalent form of farmed oceanic carnivorous fish, more recently cod and blue fin tuna are also being farmed by the Norwegians and Australians respectively.*

the production of antibiotic resistant microbes. The appearance of resistant microbes can render these important therapeutic agents ineffective for use in the human population. The fish that are used in fish farming are specially bred for commercially desirable traits. The traits that are being fostered in these fish reduces their viability in a wild environment. They are not the same as wild fish. When the fish farming industry was becoming established they were questioned about the potential of farmed fish escaping and genetically contaminating wild stocks. The industry responses indicated that it would not be a problem, but guess what, the wild fish populations are being genetically contaminated by escapees. Recently Atlantic salmon farms were started in the Pacific Ocean with the same assurances, guess what, these fish escaped too. [59] Another problem is that carnivorous fish require other fish to eat, for example, it takes 3 kg of food to produce 1 kg of salmon.

Now people are fishing for the smaller fish to feed to the farmed fish further depleting food sources both for human beings and the wild fish populations. Much of the small fish such as sardines and pilchards that are used to produce fish meal for the carnivorous farmed fish is currently caught off the west coast of Africa. The people who live in these areas are desperately poor and have relied upon these small fish as an important dietary resource. The African fishermen use traditional small boat fishing methods that can not compete with large state of the art ships equipped with the latest fish location and capturing technology. Thus, a local resource that was sustainably fished and

---

[59]   *It should be noted that the Atlantic salmon are much different than the Pacific salmon, it is unknown what the effects the introduction of these foreign fish will produce.*

provided much needed food in Africa has been co-opted to provide an inefficient source of food for upscale western and Japanese consumption.

Why have CAFO's made their appearance in recent decades when traditional farming practices have existed for thousands of years without them? The reason is that it is supported by subsidies that favor large industrial agribusiness. For example, much of the cattle in the US is finished in feed lots (generally smaller animals are purchased from starter operations) where they are fed food principally composed of field corn which is also a subsidized crop. Excessive amounts of corn is not good for cattle since their digestive systems are adapted to eating grasses (they can only survive on a diet of corn for about 3 months), however, it does make them gain weight rapidly. As a result of over crowding and the type of diet used in these facilities they have a high rate of illnesses. In order to keep them from succumbing to these illnesses their feed is laced with antibiotics, again medical experts are warning that this will produce antibiotic resistant strains of microbes. These operations are also the beneficiaries of petroleum subsidies, it takes 284 gallons of oil to support a cow that is raised this way.[xlii] We went from having a system of animal husbandry where the cow would be left out to graze by themselves often times on marginal grassland unsuited to crop production, to one that consumes huge amounts of food produced on prime agricultural land (most of the crops grown in the US are used for animal feed). Essentially what has been accomplished is a switch from a solar powered food system to a fossil fuel powered means of food production that increases pollutants and health risks!

Mining: In general mining is characterized by extraction of the nonrenewable resources for metal ore, coal, gems, and salts.[60] The process of mining usually produces large excavations, toxic byproducts such as mining tailings and processing agents which are frequently stored in retention basins similar to the ones described for CAFO's. These residues of the mining process generally produce large scale environmental degradation that is costly to mitigate. As much of these costs as possible are externalized by the owners to fall upon tax payers and also members of the surrounding communities who often have to live and deal with the engendered environmental problems. For example, if hard rock mining for minerals are considered, in the United States the owners of the older mines are long gone leaving the entire clean up costs to the taxpayer. More recently the government has required these companies to post bonds for anticipated clean up costs, however, the assessment of the clean up costs are left to the mining companies. At the present time, the clean up bonds are generally under funded by 1 ½ to 2 times and 10 times if acid[61] is being produced by the percolation of water through the defunct mine. The current tab that US taxpayers will have to pay is $12 billion for the clean up of mine residues.

The strategy the mine owners have employed to evade clean up costs is to award themselves with dividends in

---

[60]   *If the final condition of nonrenewable depletion of the mined resource is considered, this concept can be extended to other resources such as soil, plant, animal and marine organisms, for example, the passenger pigeon was hunted to extinction and was "mined out" or the desertification of a once productive area.*

[61]   *Sulfuric acid (battery acid) is a strong acid which is formed in the mine and drains out into the water shed producing severe environmental problems such as large kills of aquatic life.*

order to remove any accessible cash, shift any other valuable assets into other businesses they own, and then declare bankruptcy. For example, Galactic Resources located in Colorado was gold mining using a method to produce gold by employing the heap–leach technique which uses cyanide to extract the gold from the ore. The highly toxic byproducts this process produces including cyanide residues are stored in retention basins. According to the estimate made by Galactic Resources for the cost of clean-up they provided a $4.5 million clean–up bond. Eventually their heap-leach retention basin overflowed causing wide spread damage to the surrounding areas. The cost to the Government was $180 million for the clean up. The government sued Galactic Resources who declared bankruptcy in 1992 following the pattern described above. The Government did manage to recover an additional $28 million from the bankruptcy, leaving $147.5 million for the taxpayer to pay and a degraded environment for the surrounding communities.[xliii] The Indians in the United States have emerged as a formidable counter force to forestall this type of exploitation by the multinational mining industry. In 1975 Exxon tried to establish the Crandon zinc – copper mine using the heap-leach method in Wisconsin near the Chippewa Indian reservation which is located a mile down stream on the Wolf river, one of the few remaining pristine rivers. The river is the source of water needed for the growth of wild rice, which is an important source of food and also for the Chippewa's cultural activities. The Chippewa resisted the establishment of the Crandon mine and Exxon shelved the project in the mid 1980's. In 1993 Exxon returned along with their Canadian based Mining partner Rio Algon to establish the mine. However, by 1993 the Chippewa along with the other nearby Indian tribes (Menominee, Pottawatomie, and Mohican) had become much more savvy when dealing with the government and news media as a result of the Indian rights movement. They also had established tribe owned gambling establishments which provided the financial resources to contest the issue. While the mining case was wending it's way through the courts, the Indians received clarification on their treaty rights which granted them the power to classify water and air purity according to EPA standards. [62] The tribes promptly classified these resources to the high EPA purity levels which precluded the establishment of a mine upstream from the reservation which would have degraded their water to a lower standard. They also formed alliances with nearby non-tribal communities, environmental organizations, unions, and the sport fishing groups who realized that if the mine became operational it would have severe negative impacts upon the fisheries. Exxon/Algon tried to outmaneuver the local opposition by hatching a secret deal in 1996 to allow mining with the town board of nearby Nashville which was illegal.[63] They also sought to influence the state legislature by utilizing the largest business lobby and hiring James Klauser a well connected former Exxon lobbyist to undermine the tribes treaty rights. Another maneuver they tried was to offer the tribal leadership money to "lease" the tribes treaty rights. The Nashville Town board was replaced in 1997 and the mining agreement was rescinded in 1998. This wasn't the end though; the legal battles continued until 28 October 2003 and the courts found in favor of the Indian Tribes. The Chippewa and Pottawatomie Indian tribes also jointly

---

[62]   *The treaties between the US Government and the Native Americans does not grant mineral rights to the tribes.*

[63]   Closed meetings are illegal.

purchased the 5000 acre mining site for $16.5 million and divided ownership between the two tribes to preclude any future mining on the site concluding the 28 year struggle.

Coal mines in the United States are currently producing around1 1/8 billion tonnes of coal each year, almost all of which is consumed to produce electrical power.[64]

The combustion of the coal produces a number of toxic waste products and fly ash. Mining of coal is also highly destructive to the environment. It produces large scale environmental disruption from the excavation process, tailings, and large quantities of the greenhouse gas methane.

The pollutants produced by coal are comprised of acids, particulate matter, heavy metals such as mercury, selenium, cadmium, lead, arsenic, and the gasses carbon dioxide, methane, sulfur dioxide and nitrogen oxides. The sulfur dioxide after it enters the atmosphere is comprised of particulates and also is transformed into sulfuric acid by coming into contact with the moisture and oxygen present in the air. The air born Nitrogen oxides combine with volatile organic compounds to form particulates, smog, and ozone. In the 1950's researchers established the linkage between coal combustion emissions and increased rates of mortality which from that time until several decades ago was about 50,000 deaths / year with an additional 5,000 deaths /year as a result of the inhalation of

---

[64]　　In 2005 coal produced 40 % of all electricity produced world wide (7,344 TWh, a TWh is 1trillion watt hours of electricity). To produce this electricity 7,856 billion tonnes of CO_2 was produced, 30% of CO_2 emissions. An additional 3,124 billion tonnes of CO_2 were produced by its use in industry, transportation,space heating, and agriculture bringing the total to 41 % of CO_2 emissions world wide.

particles in mines which produced the incurable and often fatal black lung disease. According to a paper published in 2011 by Epstein et. al., of the Harvard Medical School Center for Health and the Global Environment[xliv] the number of fatalities has dropped to around 24,000 / year[65] as a result of the addition of 105 scrubbers. In this paper the externalized costs being picked up by the public are estimated and totaled then added to the cost of electricity to arrive at what the actual costs / kilo watt hour (kWh) is. Since this is the latest comprehensive peer reviewed body of information currently available I will quote their summary of the findings:[66]

Category..Estimated Costs in 2008 (USD)

Land Disturbance Carbon & Methane.....................$2.2 Billion

Public Health Burden in

Appalachian Communities.......................................$74.6 Billion

Fatalities Among the Public due

---

[65]   When referring to the Harvard study which provides a low, mid , and high values, the mid range values will be used which they consider to be the most likely. In this example the range was from 13,000 – 34,000 deaths / year.

[66]   This list is quoted from the executive summary of Mining Coal Mounting Costs (see full citation in the bibliography), on page 3, Internet source.

to Coal transport by RAIL..$1.8
Billion

Emission of Air Pollutants from Combustion............$187.5
Billion

Mercury Impacts...$5.5
Billion

Subsidies...$3.2
Billion

Abandoned Mine lands..$8.8
Billion

Climate Contribution from Combustion.....................$61.7
Billion

Total: $345 Billion

If these costs were added to the amount currently being charged for electricity it would increase the cost of electrical power by $0.18 per kWh. For example, if you are currently paying $0.08 per kWh the true cost of electricity would be $0.26 per kWh with out the externalized costs which we are indirectly paying anyhow, **coal is not a low cost source of energy but one of the most expensive and by far the most destructive.**

If a closer look is taken of the coal engendered air born pollutants, not only are they causing land based problems but also in the oceans. The combustion products produce several sources of acid, carbonic acid from dissolved carbon dioxide, and acid that is formed from sulfur dioxide which produces acid rain. The acid rain has made some of the lakes on the east coast inhospitable to aquatic life. It also produces crop damage reducing yields, and has killed large swaths of forest and degraded even more. It also eats into, degrades, and destroys infrastructure, buildings, machinery, works of art, textiles, and of course our health. Recent evidence suggests the current EPA standards are inadequate and that a further reduction of at least 80% of the sulfur dioxide emissions are needed above the 1990 standards. At the present time 81 million people live in areas of the North East United States and Eastern Canada that fall below the current EPA standards.[xlv] If coal combustion pollutants impact on human health is considered, the result of the sulfur dioxide that is produced gives rise to respiratory illnesses. These illnesses include nasal congestion, shortness of breath, asthma attacks, pulmonary inflammation, heart arrhythmia, and greater numbers of infant deaths. Coal combustion also produces additional pollutants besides sulfur oxides such as particulates and nitrogen oxides. Exposure to particulates are one of the major sources of heart attacks and lung cancer. The nitrogen oxides produce smog an irritant, and ozone which attacks the lining of the lungs.

Coal mining is also a highly destructive enterprise, the methods used are tunneling and open pit. At the present time the open pit method is the most prevalent technique. It typically requires the removal of overburden (the material laying on top of the coal seam) which is blasted away with massive explosions and then removed. The most common open pit practice is to blast off entire mountain tops (MTR) and then push the rubble off the mine into the valleys where streams are often found, thereby producing severe ecological problems. Once the mine is in operation it produces tailings which are deposited in retention basins which fail occasionally inundating surrounding areas. The mining byproducts also seep into the underground aquifers that the people are reliant on for potable water produced by wells tapping the aquifers. In mining areas such as West Virginia much of the water has already been contaminated and rendered unusable by the heavy metals arsenic, selenium, cadmium, beryllium, barium, antimony, thallium and lead. At the present time there is no way of removing mining pollutants to restore aquifers. It will require ages for natural processes to accomplish restoration. Some of the tunnel type mines have also caught on fire which generally smolder on for decades forming hollows under the ground in unknown locations. These hollows often suddenly collapse without warning producing large holes and of course lots of air pollution. Coal mining also produces large quantities of methane which is a greenhouse gas 25 five times more potent than carbon dioxide and has the same effect on the environment as the emission of 71,100,000 tonnes of carbon dioxide.

~~~~~~~~~~~

Chapter 4
Red Herrings

Even when the plum has wilted and winter has reached its deepest cold,

do not let your body be numb or your mind absent.

Dogen Zenji

Shortly after George W. Bush was elected to his first term of the United States presidency, as is customary, he gave a state of the union address which I watched on TV. During this speech he mentioned that it was being anticipated that ethanol alcohol production from corn would enjoy continuing support by the government, and was expected to become a significantly more prominent source of fuel. I wondered how this could be? When I was a university student my major was in chemistry [67] and I knew that ethanol alcohol did not have a high energy content. As a result of this curiosity I started to investigate alcohol fuel. After evaluating grain alcohol's fuel potential, I expanded the appraisal to other sources of energy that are often touted as providing significant potential to replace the current fossil fuel based economy.

In order to evaluate the feasibility of replacing fossil fuels they have to be compared to the potential that alternatives offer which require numerical values. The format that I will use will be to: alert you that an evaluation is going to be made; give some background information so that the answer(s) will make sense, provide the arithmetic proofs as

---

[67]   *My area of concentration was in organic chemistry and I also took a course in nuclear and radio chemistry.*

a foot note so that you can see how the conclusions were arrived at if you choose, and give the conclusion sandwiched between a top and bottom centered divider like this -----0-----. That way you can bypass the arithmetic if you wish and just read the conclusions. To start, the amount of potential alcohol has as an energy source to replace the oil used for transportation will be evaluated. In order to examine this question a few preliminary facts need to be known. In 2013 the daily consumption of oil in the United States was 18.89 million barrels each day. The continental United States has 340 million acres of arable land. 1 acre produces 140 bushels of corn [68], which yields 392 gallons of alcohol. Alcohol has less energy content than oil (52%), thus, 1.923 gallons of alcohol has the same amount of energy as 1 gallon of oil.[69]

---

[68] *The yield of crops varies from year to year in 2012 the yield for corn was 123.4 bushels/acre and in 2013 158.8 bushels/acre, the average for those two years is 141.1 bushels. The average for the last 30 years was 137 bushels per acre during years of average rainfall. I will use 140 bushels per acre.*

[69] *So 1 acre of land can produce the equivalent energy of 203.84 gallons of oil, i.e., 392 gallons of alcohol / acre X .52 = 203.84 gallons of oil.*

1 barrel contains 42 gallons of oil. So 1 acre produces the equivalent energy of; 203.84 gals. Oil / 42 gal. / barrel = 4.853 barrels of oil/ acre.

In a year (1 growing season) we use 18.89 million barrels of oil / day X 354.25 days /year = 6,880,682,500 barrels of oil per year.

If we used all of the crop land in the continental United states to grow corn for alcohol production we get 340 million acres X 4.853 barrels / acre = 1,650,020,000 barrels each year.

If we use the entire arable agricultural land to produce alcohol in the continental United States it would power our transportation fleet for **2.9 months**.

-----0-----

There are a few other problems though; it requires fossil fuel to produce the agricultural chemicals, till the fields, sow the crops, harvest the crops, and then process it into alcohol. The amount of energy to do all this is controversial some studies indicate it uses more fossil fuel energy units than you get from the alcohol and others less. An EIA study indicated that for every unit of energy available in the alcohol .72 units of fossil fuel energy was required to produce it, a positive energy balance. Pimental did the same type of study and found that producing alcohol from corn had a negative energy balance, 1 unit of fossil fuel input energy produced .778 units of alcohol energy. The difference between the Pimental study and the government study arises from the fact that Pimental included the amount of energy required to mine, process, transport, and fabricate the alcohol production facility. Pimental is right to include this additional energy since if the alcohol where not being produced no production plant would be needed. Moreover, the other types of crop based bio-fuels share the same problems as alcohol. Pimental and Patzek found that 1 unit of fossil fuel input energy produces .668 units of energy from switch grass and .534 units for

---

So if the entire US agricultural production were used to make alcohol it could offset 1,650,020,000 / 6,880,682,500 = .24 X 100 = 24% of our current oil consumption for transportation!

Or to put it another way it would provide transportation fuel for 2.9 months!

soybean bio-diesel. Another problem is that conventional agricultural methods of crop production also produce methane and nitrous oxide which are powerful greenhouse gasses.

From the above analysis **it's obvious that alcohol and the other crop produced bio-fuels which share the same limitations, are not able to come even close to providing the energy needed to replace a significant amount of the fossil fuels** that are currently being consumed. Don't forget that this analysis was only for transportation oil and did not include coal!

Much is said about the "hydrogen economy" but the people who are promoting it ignore some basic facts. Hydrogen **is a potential way of storing energy. It is not a source of energy.** To obtain hydrogen it must be produced using other sources of energy, and the amount of energy used to produce it is greater than the amount it can store. At the present time most of the hydrogen is produced by a process called steam reforming which combines methane (a fossil fuel) with water. This process produces carbon monoxide and hydrogen. It requires 1.25 units of production energy (from fossil fuel) to produce 1 unit of the energy content in the hydrogen; a large energy loss. The carbon monoxide is a weak greenhouse gas but its greenhouse effect is large because it prevents methane, a powerful greenhouse gas from breaking down, and also produces ozone. A similar production method called water gas uses coal and water to produce carbon dioxide and hydrogen. The water gas method produces one molecule of carbon dioxide for each molecule of hydrogen. This is a lot of carbon dioxide. Now let's consider rarely used systems that are being researched. Plasma can be used to break down hydrocarbons (such as some of our waste; plastic for example), however, this system is very inefficient, it requires 2.08 units of input energy to produce 1 unit of hydrogen energy. At the present time a few small scale commercial plasma units operating, mostly for waste product processing.

Experimental and micro scale hydrogen production methods. Thermal chemical methods: sulfur / iodine uses 2 units input energy to 1 hydrogen and copper / chlorine, 2 1/3 units input energy to 1 energy unit of hydrogen. Direct electrical production approximately 1.8 units input energy to 1 hydrogen energy unit. A number of biological systems are also being researched. These examples were chosen as a representative cross section of the types of research being conducted and there are quite a few more systems being worked on.

Let's consider the first group which is being currently used to produce hydrogen. They all produce green house gasses (which we are trying to eliminate) and also use fossil fuels to produce. They also store only a fraction of the energy used for production, thus, hydrogen as an energy storage system produced this way is a lot less efficient and expensive than just using current conventional energy sources.

Systems that uses electricity as an energy source to produce hydrogen. Plasma can be used to produce hydrogen from some of our waste but uses lots of electrical energy. Direct electrical production from water, is another system that is inefficient and expensive, it uses 1.8 times the amount of energy to produce as can be stored in the hydrogen. A further problem is that at the present time about 2/3 of the electrical energy in the United States is produced by fossil fuels that generate green house gasses. Unless the electrical energy to power these processes come from non-fossil fuel energy sources they will produce more air pollutants from the fossil fuels burnt in the power plants than using conventional gas or diesel engines.

Biological systems - Some are in the lab and others theoretical, they have the same disadvantages as alcohol and other bio fuels, requiring grown or harvested feed materials as described above.

One often hears much about the "potential" of experimental processes that will soon solve the hydrogen production problem. The question is how likely are these experimental processes to produce a viable product? I have a significant amount of experience in this area, having worked in commercial research and development (R&D) and also the laboratory. Here is how the process works:

1. One has an idea that can theoretically produce an improvement in something or for an entirely new product.

2. An experimental lab design is devised to perform simple test(s) of the idea. In many cases the idea is dropped at this stage because the results are insufficient to warrant further investment of resources. However, if the idea does show some promise it will either be pursued further or moved to stage 4.

3. If the invention is pursued further, various improvements will be attempted which will either produce a positive enough result to keep working on it or it will be dropped. If it isn't dropped and still seems promising,

4. the invention is then embodied in a prototype where it is further debugged or refined until it works OK. At this stage again many inventions are dropped because they have some flaw that can't be corrected. If the invention succeeds in making it through this stage it may be patented if warranted.

5. The last part is scaling up for a production model, again many inventions can not be scaled up for viable commercial production. This is the last stage where an invention can be patented. Once it becomes known about, it enters the public domain and can not be patented. Then

the product is placed in commercial production and may or may not succeed.

If you consider the process just described, the promising idea has to successfully make it through many stages, as a result very few ideas ever actually end up as a useful product. Even if the new device or process makes it through the last stage very few of them are actually successful. At the present time the United States Patent & Trademark office has about 9 million patents on file. At the present time over 300,000 patents are currently being granted each year. Obviously a very minute number of these ideas ever turn out to be actually utilized in a successful commercial product. So when you here about something being worked on in a lab or is under development being promoted as being "near" to solving some problem with just some more development needed. I would suggest that what is being said should be considered with skepticism. If the status of hydrogen production is considered, the likelihood that any of these lab projects actually ending up with a viable product is very small to practically nonexistent. Particularly if the device being touted is starting out with the handicap of having the inherent deficiencies that was pointed out above, i.e., poor energy balance, environmental problems, high expense, etc.

The second great hurdle that a hydrogen economy must overcome is storage of the hydrogen. At the present time a few ways are available for storing hydrogen. The current methods of storage are in high pressure tanks and in special metal alloys. In tanks, currently hydrogen is stored at pressures ranging from 340 to 680 atmospheres (5,000 – 10,000 psi). These numbers suggest that a lot of hydrogen can be squeezed into a tank. However, hydrogen is the smallest and lightest gas molecule which means that the actual weight of hydrogen is small. For example if a 44.8 liter tank (about 10 gallons) which is around the size found on a small or mid sized car were filled with hydrogen at the high pressure of 680 atmospheres it would contain 2.72 kg (5.98 pounds) of hydrogen, which isn't enough fuel to provide adequate cruising range for a vehicle. Liquid hydrogen is not practicable for use either. It must be maintained at extremely low temperatures (-253 C or -423 F) to be stabile, otherwise it changes to a gas and must be vented. Compressing and liquifying hydrogen also requires large quantities of energy to accomplish producing another unfavorable loss of energy. The metal alloy adsorption type of hydrogen storage is very heavy, extremely expensive, and has impractical charging and discharging requirements. At the present time there are about 20 different approaches to these problems that are either theoretically possible or being worked on in the laboratory. None of the ones that are being researched in the lab are ready to be moved to the prototype stage of development. So the hydrogen storage problem isn't even close to being solved.

Hydrogen also has a number of other problems. Since it is such a tiny molecule it is very hard to contain, it escapes from piping joints, valves etc. Thus, the current types of "plumbing" systems are specially made and also require high maintenance making them expensive. It also burns with a very hot transparent flame that can not be seen which makes hydrogen fires particularly dangerous.

To sum up, in order for a hydrogen fuel economy to become practical it would require a revolutionary improvement in production and another revolutionary improvement in storage. Each one of these hurdles is about the same order of magnitude as the change from vacuum tubes in electronics to semiconductors. It would also require incremental improvements in the plumbing systems and safety mechanisms for general usage in vehicles. In my view the likelihood that all of these problems will be solved within a time frame that would produce a workable system that is   practical to solve our environmental and other fossil fuel problems is remote. It will almost certainly be decades before there is even a chance that this type of technology can be implemented.

Solar Voltaic - Electricity generated from solar energy has one insurmountable problem **it only produces electricity when the sun shines on the solar cell.** Not only does it require sunlight, but it also only produces electric output near its rating for "peak sunlight' for 4 1/2 – 5 1/2 hours in the middle of the day. You won't get energy from solar cells on cloudy days or if they are covered up with snow, ice, leaves, dirt or other sun blockers. At the present time there is no cost effective way of storing large quantities of electricity to mitigate this problem (that's why hydrogen fuel cells are being researched). It can be stored in batteries but their cost is prohibitive for storage of large amounts of power. To demonstrate this problem, where I live 54% of the days are overcast. Last winter it seldom got above freezing, and many of the days were below -27 C (0 F) with some days as low as -35 C (-25 F). All the heating systems except hand stoked stoves and fireplaces are operated by electrical systems, that means for less than half the winter days the heating system where I live would have worked for maybe 5 hours a day, as well as the refrigerator and other electrical appliances. Obviously this level of performance isn't acceptable for just about anyone living in the United States.

Lets consider if solar voltaic may be practical for reducing fossil fuel use by operating it for the 5 hours (this average will be used henceforth) on sun lit days to partially offset fossil fuel use. To evaluate this aspect of solar voltaic we need to consider the amount of energy needed which according to extrapolations of Epsteins figures are a colossal 2.061 trillion kWh (kilowatt hours) per year. The solar cells used for large scale arrays are 15% efficient and have an average cost of $3,250.00 per kWh.

The amount of energy in peak sunlight is 1 kWh per square meter. I will assume that peak sunlight will be available 80% of the year with high air clarity.[70]

---

[70]   *To convert our yearly usage into hourly usage (kWh) we need to take 2,061 Billion kWh / (364.24 days /year X 24 hr/day) = 235,758,407 kWh I will round to 236,000,000 for the following calculations. (Note; not all electricity is produced by fossil fuels).*

The output of a solar cell is reduced by about 25% through conversion to alternating current, steeping up losses, line losses etc. So we can expect to get .15 X 1 kWh/square meter X .75 = .112 kWh per square meter.

The number of 1 square meter panels we need are: 236,000,000 kWh / .112 kWh / sq. m = 2.11 billion square meters.
    It requires 1 / .112 kWh/ sq. m = 8.93 square meters of solar cells to produce 1 kWh.

In order to produce the 236,000,000 kWh of energy we need : 236,000,000 X 8.93 panels = 2.11 Billion panels

        The cost of the solar cell panels is: 236,000,000 kWh X $3250.00 /kWh = $767 Billion.

        The amount of land required: An acre of land can accommodate an array of: 29 rows each having 63 square meter panels or: 29 rows X 63 panels = 1,827 panels / acre.

        So 2,110,000,000 panels / 1,827 panels/ acre = 1.15 million acres

        According to the FHA the average cost of land in 2010 was $1,000 / acre

    The land cost is $1.15 Billion

        In order to operate this system the power transmission grid would need to be upgraded (the smart grid) at a cost of $1.2 Trillion.

    The total for this system is; $767 billion + $1.15 billion + $1.2 trillion = $1.97 Trillion.

-----0-----

The Total Cost for solar voltaic is $1.97 Trillion, WOW! It would also require

1,150,000 acres of land. This system would provide electric power for 16 2/3 % of the year!

-----0-----

It should be noted that the above cost does not include the wiring to connect the panels, if all the panels were lined up end to end they would stretch for 2.11 million kilometers (1.3 million miles).

The amount of wire required certainly would be stupendous with a proportionately large cost, so the $1.97 trillion cited above is much lower than the actual cost. Moreover, these costs also do not reflect the externalized costs. For example, it would require a significant increase in mining to produce all the materials required, such as copper for wire, steel for mounting brackets, hardware, and many other types of materials incorporated in other essential components. As you probably recall mining has a lot of undesirable environmental and social costs and should be avoided if possible. The manufacture of solar cells produces toxic byproducts. Fabricating approximately 2.11 billion square meters of solar cells would produce colossal amounts of toxic waste that would have to be dealt with, probably at public expense. Shading 1.15 million acres with solar cells and the required access areas (for maintenance) would almost certainly have a significant and undesirable impact on the environment and ecology. It would also take an army or several armies of people to clean and maintain the panels. Also the taxes on almost 1.15 million acres of land would also be considerable and would be added into the electric rates. If we consider the cost, solar voltaic is prohibitive, especially since it would only theoretically reduce fossil fuel plant usage by 16 2/3%. This system would also produce no reduction in the number of fossil fueled electricity plants because they would still be needed to produce energy when the SV system wasn't operating. A final problem with this is that several types of fossil fuel electric plants are used, peakers and base load. Peakers are used to provide extra power when a sudden increase in demand occurs and can be turned on and off rapidly. Base load plants are designed to operate continuously and provide most of the electric energy. Base load plants can not be shut down for a few hours and turned back on. Therefore, the use of solar isn't feasible to replace fossil fueled grid transmitted electrical generation at all! It's only really cost effective use is for producing power in remote locations where the cost would be prohibitive to run power transmission lines.[71]

The prospects for future fossil fuel availability. The oil and coal industries often tell us that there are huge reserves of their products and that theoretically they will not run out for hundreds of years. They are being rather disingenuous when they say this because the amount that it is feasible to recover is much smaller than the reserves they speak of.

Oil wells are usually depicted showing an underground lake of petroleum with a pipe connecting the well head on the surface to the underground lake of oil. Actually oil is found in a porous type of rock that has a structure that resembles a sponge, of course the rock can not be squeezed to force the oil out like a sponge. The number and size of the sponge

---

[71]   . *Yes it's true some people choose to live off the electric grid without electricity or are willing to pay the huge costs to establish a on site power system. For the vast majority of people though this isn't an option.*

like pores govern the amount of oil found in the rock, many large pores lots of oil. They have another measurement related to the spacing of the pores called permeability, this indicates how many of the pores touch each other allowing the oil to flow through the rock sponge to the well hole in the rock. If the well has large pores and high permeability (lots of oil with good flow) this is a high quality well. Usually the largest quantity of oil that can be recovered from a high quality well is about 1/4 of the reserves.

How about good old coal we are told that we have huge amounts of it, and this is definitely true. However, the amount of coal reserves that exists isn't the same as the amount that is economically recoverable. A 1993 study by Rohrbacher (U.S. Bureau of Mines) found that between 5% and 20% of the coal is actually economically recoverable. Since we have been mining coal for hundreds of years, most of the coal that was inexpensive to mine has already been mined.

According to this study in about 10 years the cost of coal will probably start to go up because of the increased cost to mine it. For example, the largest deposit and mining operation is in the Powder River basin where 40% of our coal is mined. When they started mining it around 40 year ago it was covered by about 6 meters (20 feet) of material (dirt, mudstone, and sandstone). This material must be removed before the coal can be dug out. This particular coal seam is sloping downward and they now have to remove about 75 meters (250 feet) of material. Needless to say, it's much more expensive to remove all the additional covering material. Since the deposit is sloping down and ever more covering material has to be removed, at some point the cost of recovering coal from this mine will become prohibitive (the Rohrbacher study estimated that

11% of the coal was recoverable from this deposit).[72] Another mathematical method used to predict resource depletion was developed by Marion King Hubbert (see below for a more detailed explanation). Using Hubbert's method of analysis the amount of coal mined will reach a peak in the US in 2015 and start to decline while becoming increasingly expensive until it's cost is prohibitive. A multicycle Hubbert analysis has shown that world coal production has already peaked in 2011 (Epstein p.74). Moreover, it will also require progressively more amounts of energy to extract eventually producing an unfavorable energy balance.

Marion King Hubbert was a geophysicist who worked for the Shell Oil Company and was aware of biological research on invasive species that started in the 1940's. The biologists developed mathematical models describing the dynamics of exploitation of resources by novel species. Hubbert realized that these models could be applied to the exploitation of nonrenewable resources by human beings. In 1956 he published a paper using this type of modeling predicting that oil production would peak in the US in the late 1960's – early 1970's. His prediction proved accurate, peak oil production occurred in the US in 1970. This type of model is comprised of several interrelated mathematical curves of a similar type to the commonly used curve developed by Gauss (the famous bell curve). One of the characteristics of Gaussian curves is that the right side is a mirror image of the left. So if you know the first half of the curve that describes increasing exploitation,you can also

---

[72]   *When I originally investigated this about 10 years ago coal cost about $30.00 / ton, it now costs around $41.00 / ton.*

produce the second half which shows the mirrored decline of the resource, thereby providing a means of prediction.

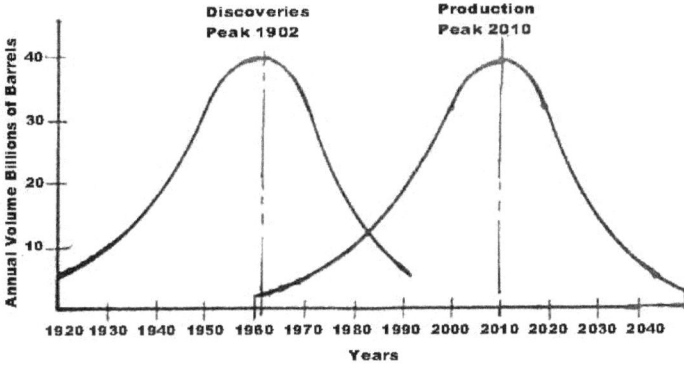

Fig. 6 Hubbert Curves for oil discoveries & depletion.

If the exploitation of oil fields are considered, initially the field is discovered by boring a test well that successfully produces oil. The next stage in the process is development of the oil field by establishing more oil wells which increases production and also provides information on its extent and quality. At some later point in time the field will be fully developed and producing its maximum output (fig. 7). While this developmental process is taking place the wells that were bored and started producing in the initial stages will become exhausted and go dry. The wells that are established progressively later in the oil fields development will also eventually go dry until the entire oil field becomes drained. So what we have are two bell curves, one that describes the developmental phase where the oil is being pumped out, followed by another similar bell curve that describes well exhaustion Fig. 6. The same type of analysis can be expanded and used to predict the peak, decline and exhaustion of world oil supplies since a fixed number of oil fields exist in the world Fig. 7.

143

Years

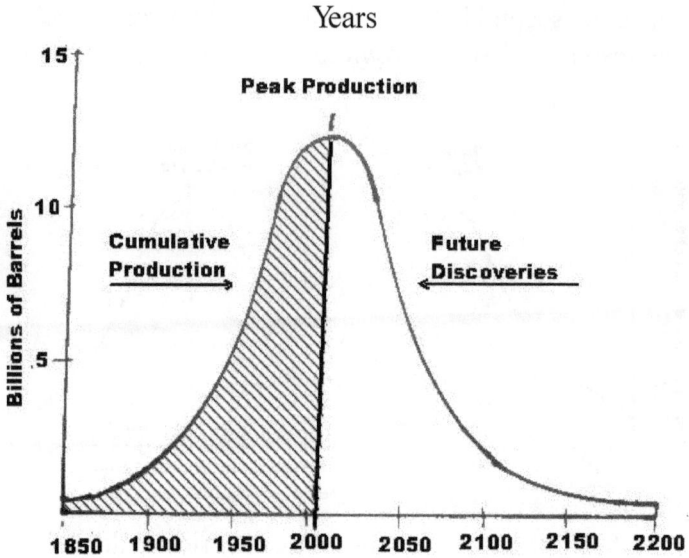

Fig. 7 Hubbert Peak for world oil production and
exhaustion.

When discussions of dwindling fossil fuel availability
comes up the fossil fuel industry quickly brings out the
magic wand of technological solutions that they claim will
continue to make their products available into the far
future. The nature and status of these wonders are usually
presented in the vaguest terms if at all, and as we have
already discussed, even inventions that are fairly well along
in the development process seldom become viable.
Another argument these industries often present is that new
resources will be found to offset the dwindling supplies. [73]

---

[73]   *If oil is considered, there is a very good reason that oil
exploration and development is mostly taking place far out in the
oceans where it is expensive and difficult. Almost all of the land
based oil has been found and there are few places where large
quantities exist that hasn't achieved a mature developmental stage.*

There is no doubt new oil and coal fields will be found, but new discoveries will become progressively less frequent as a result of the declining population of untapped fossil fuel fields. The fact of the matter is fossil fuels are finite resources that have been mined since the 19th century (oil) and coal extensively since the 17th century both in vast quantities.[74]

The coal industry also uses the argument that its product can be continued to be used because of "anticipated developments" in carbon dioxide sequestration technology. In the Harvard paper by Epstein et al., this question was considered and upon investigation it was found that to operate this type of technology large quantities of energy are required, thereby increasing the amount of coal needed to produce a kWh of energy by 25% - 40%. This of course would greatly increase coal usage with its associated environmental problems caused by mining, retention basins for tailings, fly ash, as well as the other remaining air pollution components. In this process the carbon dioxide is collected and injected into underground storage areas such as depleted oil wells. To deliver the carbon dioxide to these storage areas it would require the establishment of an extensive array of pipelines and other types of infrastructure to implement. The Norwegians had a small scale pilot project of this type it cost this plant about $200 / tonne to sequester the carbon dioxide[75], they happened to

---

[74]   *Coal was mined for hundreds of years prior to the 17th century but the industrial revolution moved it in to high gear in the 18th century!*

[75]   *It should be noted that while coal has variable composition it averages around 80% carbon. When it is trans formed into carbon dioxide 12 unit weights of carbon require 32*

have a nearby depleted off-shore oil well so the pipe line to it from the power plant was not very long. In 2006 they decided to try to scale their system up to process carbon dioxide on an industrial scale with a target date of 2020 for start up. The project was abandoned after spending $1.2 billion with an additional $290 million in cost over runs. The implementation of this type of technology also has other problems besides being prohibitively expensive and inflicting greater damage to the environment. Carbon dioxide which is odorless and tasteless becomes a deadly gas in high concentrations. If the system developed a leak either through fissures in the underground storage area or the pipelines, it has the potential of causing deadly widespread catastrophes. Of course what will be argued is that the carbon dioxide can not escape from far underground but as you will recall carbon dioxide in the presence of water forms carbonic acid which can eat through many types of rock, corrode metals, and degrade other types of materials. The presence of underground water is quite common in oil wells. In fact water is frequently injected into oil wells in order to maintain oil well pressure and/or scavenge residual oil. This is where they are planning on injecting the carbon dioxide, is this wise?

Natural gas started to be exploited for energy later than petroleum (it was usually just burnt off in the early days of oil extraction) and gas wells are also found at a greater depth which put them out of reach of the early technology. If a Hubbert type of analysis of natural gas is performed it should peak and go on the decline about 10 or 15 years

---

unit weights of oxygen, thus, 1 tonne of coal will produce approximately 2.9 tonnes of carbon dioxide.

after oil or between 2020 – 2025 since world oil peaked in 2010.

The question is, why are these industries the authors of expensive PR campaigns to deflect attention away from these unpleasant facts? The most pressing motivation is to maintain and increase their short term profit margins. If we look at it from a longer range perspective, if they can eliminate or attenuate the implementation of competing non fossil fuel energy sources the greatest profits will be made during the period when critical shortages develop. The most basic economic principal is of supply and demand, if demand for some thing is high and its availability is low, the prices the commodity will command will be high. So if you own the hole in the ground that the needed fossil fuel comes out of you are going to make huge profits! Conversely if alternative sources of energy are implemented not only will it reduce fossil fuel sales, it will also hold the unit price down. So basically what you will see are all kinds of fossil fuel industry PR designed to string us along. The themes of these PR efforts will be to utilize smear and disinformation campaigns directed against alternatives that have the capacity to significantly reduce or eliminate the use of these products. The other purpose of these PR efforts will be to create an impression that solutions to fossil fuel problems are just around the corner, and there is no need to change. Oreskes and Conway gives a detailed history of some of the PR methods employed in their excellent book Merchants of Doubt.[xlvi]

Manure digesters utilize a capped basin to ferment and capture methane produced by animal and or humane waste. Methane is an excellent fuel, however, when you burn methane it is converted to carbon dioxide. Thus, you are changing a powerful green house gas that stays in the air

147

for 12 years to a weaker one.[76] According to Archer it requires two centuries for carbon dioxide to diminish by 3/4 and an additional 10,000 years for it to fall to 10% - 12% with the remainder staying for hundreds of thousands of years.[xlvii] Because of the long period of time carbon dioxide remains in the atmosphere it causes 40,000,000 calories to be absorbed in the air for every calorie derived from its combustion. Thus, production of methane from manure should be avoided since whether you burn it or not it is extraordinarily detrimental to the environment. We will return to this topic later on.

Geothermal is another technology that is often held up as a potential means of replacing fossil fuels. At the present time it provides .41% of the energy in the United States. It taps the high temperatures found deep underground by boring two holes and fracturing the rock between the holes to make the rock permeable. The system operates by injecting water down one hole then passing the water through the hot fractured rock where it is converted to steam which goes up the other hole. When the steam arrives at the surface it is used to power a conventional steam turbine electrical generator.

A geothermal plant using this system was constructed near Bern Switzerland which had some problems. When it was in the start up phase it produced a number of minor earthquakes,[77] the largest were around 3.5 magnitude. The

---

[76] Atmospheric methane oxidizes to carbon dioxide in 12 years so whether you burn it or not it produces carbon dioxide.

[77] *This area of Switzerland had experienced an earthquake about 500 years ago.*

Swiss project was abandoned because of the earthquakes. A similar project that was underway but not as far along in the United States was abandoned shortly thereafter because of the earthquake problem. There are a few of these systems that produce significant amounts of energy that are in operation, for example, in Iceland. However, the areas where this form of energy can be effectively tapped is dependent upon rare geological formations where they can extract energy from volcanic heat stored in rocks near the earths surface. Recently the Icelanders have established a system that utilizes heat directly from the molten lava near the surface, not from very deep wells. One can't deny that a lot of subterranean energy exists, however at the present time there are no feasible systems that could be widely applied as a large scale general source of energy.

Heat pumps[78] work by either transferring heat from the subsoil into the building for heating or absorbing heat from the building releasing into the subsoil for air conditioning. They employ a modified technology similar to that used for refrigerators and air conditioning. The difference is that the system can be reversed. It is a very efficient system, it only uses 1 unit of energy for every 5 units of energy exchanged from the subsoil. Heat pumps have been in use for about 35 years and almost all of them are still in use; they are durable and reliable. They are also expensive, the transfer of energy from the building to the soil requires

---

[78] *Recently the heat pump industry has started referring to this type of heating system as "geothermal" in their advertising PR for some unknown reason. In order to avoid producing confusion in our discussion I continue to use the original designations,*

either a long trench to be dug or deep holes to be bored into the ground. Tubing is buried in the excavation to conduct heat from or to the surrounding earth. Lin found that the average cost for a heat pump system was $9,850.00 greater than conventional systems.[xlviii] He calculated that it would require 22 years to offset the additional initial heating system cost with energy savings (assuming a 5% interest rate). He also noted that 21% of the current inventory of housing would not be able to install a heat pump system because of inadequate property space. At the present time less than 1/2% of the housing in the United States uses this type of heating system. Basically the cost of this type of system is prohibitive.

There are many different inventions being worked on to utilize various forms of energy present in the Oceans. These efforts fall into several general categories based upon kinetic energy or physical properties; mechanical (kinetic) methods that utilize water pressure differences that are captured as a differential in the height of tides and turbines that use the flow of water in ocean currents or flows from changing tides to drive them. The second group uses differences in water properties such as temperature differentials between surface water which is warm and the cold abyssal depths or differences in the salinity between fresh water and ocean water. At the present time the general level of development for the more perfected of these approaches are in the small scale pilot plant phase some of which are producing a few megawatts.[79]

The majority of these inventions are in the lab or theoretical stage of development. It seems unlikely that this source of

---

[79]   *It should be kept in mind that 1 megawatt is comprised of 1000 kilowatts, a tiny amount of energy when compared to what is currently being used.*

energy will become prominent for at least a number of decades. It also has the disadvantage of being confined to the oceans. Many of these schemes have the further shortcomings of being likely to produce environmental problems since many of them rely on changing the salinity of the water, significantly altering its temperature, or altering the flow rate which can cause sedimentation and other ecological problems.

Hydroelectric at the present time is the source of 7% of the electrical power that is produced in the United States. It produces no air pollutants, provides the significant benefits of flood control and irrigation enabling large tracts of land to become agriculturally productive. It has the disadvantage of being disruptive to the ecosystem. In recent years much discussion has been taking place about removing some of the dams that are having deleterious effects on the marine environment in estuaries and coastal regions. One often hears arguments that it can be increased by installing "micro" hydroelectric plants on small streams. Installation of small scale hydroelectric dams is possible, but the problem of ecological disruption is present for these projects as well. If the potential for the amount of power is considered, micro-hydroelectricity's capability to offset fossil fuel usage is trivial. Therefor, the prospects for any significant expansion of hydroelectric power generation in the United States is some where between slim and none since this resource is currently almost completely being exploited and the ecological problems it produces are unacceptable.

Molten salt reactors are a type of nuclear fission reactor that utilize thorium for fuel. Thorium is more plentiful than uranium and has much less radioactive activity. Because of it's greater availability it has great potential for the production of cheap plentiful energy very far into the future. This type of reactor is also much more effective at extracting energy form it's fuel and therefor produces a small fraction of the nuclear waste produced by the light water nuclear reactors in current use. The nuclear waste that is produced by these fast neutron reactors diminishes to a safe level in hundreds of years instead of the hundreds of thousands of years required for the second generation reactors that are now in use.

Thorium also has the further advantage as a reactor fuel because it, and it's byproducts can not be used to make atomic bombs. It also has the further advantage of simply being a technical extension to a different fuel of an already existing technology (the fast neutron liquid metal reactor discussed later on). At the present time a number of projects are underway to develop this type of reactor. To me out of all the technologies discussed in this chapter molten salt reactors seems to have the greatest potential to eventually come to fruition.

Hydrogen fusion has had much research done over the last 40 years. It is in the lab stage of development. Every time the question is asked when a commercial system will become available the answer has been (for last 40 years) "maybe in 40 years". My guess is this technology probably will be mastered eventually and possibly become a useful energy source in the 22nd or 23rd century!

Space based solar voltaic – this proposal envisions the construction of gargantuan solar cell arrays that are in geosynchronous orbits in outer space. This system would beam the power down to earth using microwave radiation. This type of solar cell system has the advantage that it can provide power 24 / 7, but it is so far beyond our current technological capabilities it's not even possible to guess when it could be implemented, certainly not within a time horizon that would provide any solution to our current problems.

Electrical power from wind mills, at the present time they produce 4.13% of the electricity in the United States and their numbers are gradually being expanded. Wind generation has the problem of being an intermittent and variable source of power. A further deficiency is that wind speeds below about 13 km/hr (8 mi/hr) contains very little energy. To overcome the shortcoming of its intermittent and variable nature proponents argue that an interconnected large scale system could produce a constant (base load) level of electrical power. In order to accomplish this an elaborate power transmission grid like the one for solar voltaic would be required that would cost $1.2 trillion. Another problem with these systems is that their actual output of power is a fraction of what they are rated at. About 10 years ago I read a windmill engineering monograph which indicated that they produce around 25% of their rated capacity. More recently Blees provided actual performance data from several large wind farms located in California and Florida, both had an output of 21% of their rated capacity.[xlix] Even using the more optimistic value of 25% we would need to build four megawatts of capacity to produce 1 megawatt of power. The cost per fully rated installed megawatt in 2013 ranged from $1.3 - $2.2 million. If we take an average of $1.75 million and divide by their actual output of 25% it actually costs $7 million for 1 megawatt of averaged output. Another unfavorable factor is that the industry specifications indicate that windmills have a 20 - 25 year life expectancy while conventional fossil and nuclear plants have a life expectancy of 60 years. So when making comparisons to fossil fuel and nuclear plants, windmills actually cost 2.4 times more using the more optimistic 25 year value. So adjusting their cost for use over a 60 year period brings their comparative cost up to $16.8 million per megawatt. Bearing in mind the differential in life expectancy I will do the arithmetic using the 25 year life expectancy cost since reducing pollution is an overriding concern at the present time. You will also notice that a larger amount of electricity 2,622 Billion kWh of electricity /yr or 300 million kW / hr. is being used for these calculations, from 2011 EIA figures . The larger value is a result of the inclusion of electricity produced by peaker plants.[80]

~~~~~0~~~~~

The total cost: $3.3 Trillion, WOW, WOW! It would require 36 million hectares (89 million acres). Actually the amount of land needed would probably be much greater according to the Harvard climate modeling study. Moreover, this system would probably produce greater climatic impacts than doubling the current amount of carbon dioxide (see below).

~~~~~0~~~~~

---

[80]   *So we need 2,622 billion kWh / (364.25 days/ year x 24 hr/day) = 300 million kWh. of electricity.*

  300 million kWh x $1,750 / kWh x 4 (to bring it up to rated capacity) = $ 2 Trillion cost for the windmills.

  The number of 1 megawatt windmills needed is 300 million kWh / 1000 kWh per windmill x 4 = 1,200,000 windmills.

  According to the literature a windmill requires 30 hectares (74.1 acres) so we need 1,200,000 x 30 hectare each = 36 million hectares (89 million acres) to site the wind mills. The cost would be $ 89 billion.

  A windmill system would also require the advanced type of electricity transmission grid that costs $1.2 Trillion.

  So if we add all these costs up it comes to $3.3 Trillion

  The cost is prohibitive, the amount of land required is too great and it probably would have negative environmental consequences.

Recent research indicates that wide spread use of windmills would produce large negative climatic impacts. The reason is that windmills take their energy from the wind by reducing its speed. By employing wide spread deployment of this technology the overall reduction in wind speeds would result in changing the wind circulation patterns according to a recent Harvard study employing climate modeling.[1] This research indicates that the effects resulting from the changes in circulation patterns would produce an impact on the climate greater than doubling carbon dioxide. Keith and Adams also found in the same study that mesoscale wind farms (100 square km and above) diminish the winds velocity from wind shadowing to a much lower speed than is currently being used for energy estimates. The corrected lower wind velocities have the effect of reducing its actual energy potential to only a fraction of current estimates. Obviously it is not possible to replace a significant portion of the United States fossil fuel generating capacity with windmills.

------------- 0 ------------

This concludes the first part of this book which is a survey to identify some of the more prominent problems that are currently facing us. Because of the great number of relevant topics that needed to be covered, of necessity I had to greatly condense a lot of information some of which was highly nuanced. It is my hope that the condensation that was made is reasonably reflective of the original authors presentations. I would urge you to consult the bibliography at the end of this book to read some of the original materials from which this book is derived. Of course any errors are my own responsibility.

Chapter 5
Regeneration

*Gold dust is a wonderful thing to have
but when it gets into the eyes, it causes blindness.*

Chinese Proverb

The question is what can we do to correct the plethora of problems that currently face us as a result of an out of control economic system that studiously ignores the widespread negative impacts it produces? In order to solve these problems within the time horizons likely to be effective (around 15 years) in preventing a significantly degraded planetary environment it will require the rapid implementation of **already existing technologies**. The second more fundamental need is to provide a social mechanism that can act as a counter balance to eliminate the negative characteristics inherent in the large scale highly extractive economic system that is dominant

From the mid 1980's and about 20 years thereafter, I was not in favor of the expansion of nuclear power. The problem in my view wasn't the safety of the electrical generation plants but the accrual of long lived radioactive waste which didn't have any real safe effective means of disposal. My views changed because of a lack of viable alternative carbon free energy sources able to produce the amount of energy we use, and also because of the advent of new power plant designs.

Let's start by considering what we currently have. At the present time almost all of our fleet of 104 nuclear power plants were built between 1969 and 1975 [81] and are comprised of 2nd generation light water (slow neutron) type plants.

These plants were actually designed and constructed 40 – 50 years ago. Nuclear power plant design has come a long way since then. The state of the art type of plants now available are 4th generation (fast neutron) plants.[82] What is really nice about the 4th generation plants is that they produce 60 times more energy per kilogram (or pound) of fuel. The second great advantage is that they can use nuclear waste as fuel. The third great advantage is that they produce miniscule amounts of waste that will decay away in hundreds of years instead of the hundreds of thousands of years for waste produced by light water plants.

At the present time there are three sources of nuclear waste that can be used as 4th generation reactor fuel before we would have to mine more uranium. At the present time about 70,000 tonnes of nuclear waste is being stored on site at the existing nuclear power plants in temporary storage casks. We also have approximately 600,000 tonnes of depleted (U238) uranium that is produced as a byproduct of the enrichment process needed to make nuclear fuel for 2nd generation reactors. A third source of fuel is approximately 500 tonnes of weapons grade plutonium produced as a result of the SALT nuclear disarmament treaties that can be diluted and used for nuclear fuel in 4th generation reactors (dilution renders the plutonium useless for bomb making). The amount of energy that can be produced in 4th

---

[81]   *The last one 1978.*

[82]   *This type of system was developed by the US government and is currently available from GE / Hitachi and is called Prism.*

generation reactors using nuclear waste for fuel is great enough to provide the energy needs of the US for thousands of years while simultaneously eliminating nuclear waste in addition to the pollution and health problems originating from fossil fuels!

The safety of nuclear electric energy generation is the question many people are concerned with so let's take a look at what the safety record of our current fleet of plants. How does commercial nuclear electrical generation compare to current fossil fuel sources? Since nuclear electrical power generation produces no air pollutants, there are 0 deaths from this source. It has no negative effects on the environment from acid rain, smog or heavy metals and does not contribute to global warming since no carbon dioxide and other warming gasses are produced.

Nuclear power plant radiation: the average annual radiation exposure originating from our nuclear power plants is less than .02 millirems and at the fence line of a nuclear plant the maximum dose is 10 (a millirem is a measurement of the damage done to a human being by radiation). To provide a few comparisons to other common sources of exposure the average x-ray dose is 95 millirems or 4,750 times more, if you live in a brick house the bricks produce an average dose of 30 millirems or 1,500 times greater. In fact if the same radiation exposure standards for nuclear power plants were applied to coal plants they would all have to be shut down!

What about nuclear disasters such as Chernobyl and Three Mile Island (TMI). First it should be noted that Chernobyl's design is completely different than western reactors and the accident that happened to Chernobyl is impossible in a western reactor. Chernobyl was housed in a regular building instead of a western type of containment structure. A regular building was used so that weapons grade plutonium could also be produced by the reactor which requires the fuel assemblies to be removed at short time intervals usually every few weeks. Containment structures makes this process slow and very inefficient. Western reactors usually replace 1/3 of their fuel assemblies every year and a half which makes reactor grade fuel plutonium useless for weapons. Chernobyl moderated the nuclear reaction with carbon rods which are similar to charcoal. These carbon rods caught on fire when the reactor overheated and caused a chemical explosion (it was not a nuclear explosion), which destroyed part of the building and ejected radioactive material. Second generation western reactors use water to moderate the nuclear reaction because water is not a flammable substance. The actual immediate death toll at Chernobyl was 31 dead. The ultimate actual number of deaths according to the 2005 United Nations study indicates there were 4,000 cases of thyroid cancer in children of which nine died. An additional 19 people died from direct exposure to the reactor, and the total number of deaths was 3,960. This gives an average death rate of 180 each year, .025% of the number of deaths caused by fossil fuel use. Three Mile Island was the worst U.S. nuclear power plant accident and also the most severe type of accident. Most of the core melted down destroying the reactor. All of the melted material remained in the reactor vessel (incidentally the "China Syndrome" scenario is an impossibility, a complete myth). The death toll was 0, the number of people injured 0, and numerous authoritative epidemiological studies have found that the amount of cancer increase has been 0.[li] What actually happened; the containment structure worked as designed. The containment structure of a western light water power reactor consists of a reactor vessel that is 7.6 – 23 cm (3 to 9 inches) thick steel. The reactor vessel is surrounded by what is known as the primary containment which is usually another 20 cm (8 inches) of steel. The two heavy steel containments are housed in a further heavily steel reinforced concrete building 1 meter (3 to 4 feet) thick. For comparison, the armored decks of battle ships are 12.7 cm (5 inches) thick steel or about 1/3 of a nuclear reactor. The outer building employs a type of concrete construction similar to what is used for hardened military fortifications. More recently on 11 March 2011 Fukushima Japan suffered a 9 magnitude earthquake which produced a tsunami that over topped the 5 meter high protective enclosure which inundated the nuclear reactors causing great damage and large releases of radioactive materials. A recent World Health Organization investigation found no increases in the rates of cancer in the effected areas. There were two cases of acute radiation poisonings of plant workers and no deaths related to radiation releases from this disaster. The contamination from releases of radioactive materials are currently below detectable limits (in 2014) as a result of the short half-lives of some of the material and also as a result of dilution.

Obviously the information on nuclear power plant safety that was just presented is almost the opposite of what is normally found in the popular mass media outlets (I urge you to check the information presented if you have any doubts). How one can distinguish true information from false is by reading **independently peer reviewed sources of information such as journals.** If you are using books or other types of media such as the internet, look at their sources. If you have someone presenting information and they are claiming to be an expert make sure they actually are expert in the area they are talking about. A sure sign that you are probably getting fed PR is when a Medical Doctor who is a specialist in obstetrics is claiming to be an expert in the medical effects of nonclinical radiation exposure, or a particle physicist is claiming to be an expert in epidemiology, etc. The large well moneyed enterprises routinely hire credentialed people like these to make PR assertions to cause doubts and confusion about real valid scientific findings that may have a negative impact on their business activities. Not only do they hire these people to create bogus information, but they have also developed an infrastructure comprised of various organizations to provide a patina of credibility to their PR. Some of these employ people to provide favorable (not independent) "peer review" and publication venues. Other organizations act to produce a provenance of publication history in order to work these articles into the more reputable mass publications by stages.[83] The general content of this type of commercial disinformation presents a supportive case for a special interest and is at odds with the general consensus shared by experts in the field. As for myself, **I am not an expert in nuclear physics, nuclear engineering, the effects of radiation exposure or epidemiology.** When I was a university student studying chemistry I did take one course in nuclear and radio chemistry which provided a basic understanding of the science underlying nuclear power. By having a basic understanding of these processes it has enabled me to identify at a minimum the more obvious PR, such as the scaremongering about "the China Syndrome" scenario which is an impossibility. Another myth is that nothing can be done to reduce our nuclear power plant waste products which actually can be reprocessed and reused in our current inventory of light water reactors for fuel.

---

[83]   *The major political parties particularly the Republican Party uses an identical system to push their agendas. In the 1980's the fundamentalist Christians, right wing, and conservatives adopted these methods as a strategy to further their interests.*

By reprocessing it is possible to eliminate most of our inventory of this material.[lii] At the present time all the other major users of nuclear power do in fact reprocess their fuel. We used to reprocess our fuel but the Carter administration stopped it for political purposes. There is no doubt that producing and having to store long lived radioactive waste is a big problem. It is strange to me though, that the antinuclear power advocates aren't urging that these wastes be sent to Canada for reprocessing (the Canadians have a reprocessing facility), or restart our own reprocessing. Reprocessing would eliminate ~ 60 % of it and there also would be no need to mine uranium for several decades.

Another source of incorrect information is the use of obsolete facts. Much of the material presented in the mass media originated in the 1970's which became "conventional wisdom" when little was known about low levels of radiation exposure, and also when nuclear power plant technology was having start up problems. Starting in the late 1960's the electric utilities began to place a lot of orders for nuclear power plants, however at that time the generating capacity of a nuclear reactor was around 200 megawatts, the utilities wanted 800 – 1000 megawatt plants. The power plant vendors responded to this demand and scaled up their designs to plants of the desired size. The degree of size scale up was beyond what the level of engineering capability could support, i.e., the scale up was too large. The result was that a lot of plants were built that had problems and needed shut downs to be debugged. To correct these problems they were off line a lot and only producing around 50% - 60% of their rated capacity. The nuclear electric generation industry tried to cover these problems up, which of course was eventually found out. When these coverups became known it caused a loss of their credibility and established a negative reputation that still lingers on.

Another source of information distortion originates from a small but very vocal group of ideologically motivated people. These people create and use controversy for their personal purposes. For example, in 1964 an ambitious Ralph Nader testified before congress that nuclear power was not needed because by the mid 1980's geothermal electrical generation would be providing most of our energy. Well, here we are 50 years later and its supplying .41% of our energy. It did serve his purposes though to provide career leverage to obtain a high position in the federal bureaucracy.

The question is, why is nuclear electric power generation still getting such bad press? It's been quietly perking along for 40 years without any significant problems since 3 Mile Island. Little is mentioned in the popular press that it has been offsetting the production of billions of tons of pollutants? Well obviously we have the opposition groups already mentioned and not all the people who oppose it are opportunists or motivated by ideology, but by concerns about its use. There is no doubt that many of the people are well intentioned but badly informed as a result of the type of the coverage it gets in the electronic and print media. For example, I recently read an article in the New York Times about a water leak at a nuclear power plant. A water pipe on the side of the cooling tower had leaked water on the side of the tower as a result of a wood support structure decaying, allowing a pipe to move out of position. The water that was being spilled on the side of the tower never had any contact with the nuclear reactor. How often have you seen this type of trivial incident reported in one of the largest newspapers in the United States if it had happened at a coal or gas fired power plant? Another interesting fact is that nuclear reactors are being built right now in the

United States for use in naval vessels. For example, the Gerald Ford aircraft carrier mentioned earlier. I would argue that a significant nuclear reactor accident is much more likely on a military ship than at a land based power plant; in fact a number of nuclear powered ships have sunk. One scarcely hears even a peep about the construction of these nuclear reactors or any protests about their deployment in a setting that has a much greater likelihood of having catastrophic results than a commercial unit.

Let's consider another possible source of opposition to nuclear power by looking at who's ox would be gored. I would speculate that much of the opposition is a result of the activities of the fossil fuel and their support industries. Once again we will use coal for our analysis keeping in mind that gas and oil are also used to produce a minor amount of electricity.[84] The amount of coal used for domestic consumption in the United States is slightly over 1 billion tons at the present time. The cost of coal is slightly over $41 per ton. Almost all of this coal is shipped by rail at a cost of a little over $17 per ton. So if we replaced coal as an energy source the owners of the coal mines would lose the profits on $41 billion and the owners of the railroads the profits on $17 billion each year. There are a number of other businesses that profit from this, for example, manufactures of explosives, mining equipment, etc. Let's consider the activities of these businesses that we do know about. The fossil fuel industry has waged an expensive, protracted PR campaign for decades to create doubt and confusion in the public mind about the scientific

---

[84]   *The cost of producing electricity using oil and gas is much greater than coal, that's why it's used for peaker plants which are only operated to adjust electric supply to intermittent fluctuations in electrical demand (they can be started up fast).*

evidence relating to the connection between greenhouse gas emissions and global warming. They have done this by employing the same type of disinformation campaign pioneered by the tobacco industry. The question is, why would they do this? The answer is that once a public consensus is reached that green house gas emissions are a cause of warming that will significantly destabilize the earths climate a motivation for change will exist. Then the appropriate action to mitigate the problem would be to reduce and ultimately eliminate fossil fuel usage. The details of their disinformation campaign are well detailed and documented in Orekes & Conway's book already sited in the bibliography. The fossil fuel industry is capable of and does perform the sorts of assessments that were performed in the last chapter of potentially or currently competitive technologies to their products. Once a hard look is taken of all these technologies it becomes obvious that nuclear power is really the only option available that can be currently implemented to significantly reduce or eliminate fossil fuel usage. If the actual performance of our light water reactors are evaluated, they have been providing 19.8% of our electrical power for the last 40 years, and at a lower cost than other methods of electric production. For example, a 1982 study by Commonwealth Edison found the following: a cost of 2.24 cents per kWh for nuclear generated electricity, and a cost of 4.33 cents per kWh for coal, coal cost 1.9 times more than nuclear. A slightly later study by the European Economic Community produced a study that showed similar results. A more recent 2005 analysis of the comparative production costs of energy sources for electrical generation gave the following values per kWh: nuclear 1.72 cents, coal 2.12 cents, natural gas 7.5 cents, and oil 8.09 cents.[liii]

Light water reactors do have several disadvantages. The largest disadvantage is that they produce large amounts of long lived radioactive waste. The second disadvantage is that they are expensive around $4000.00 per kWh to construct. Upon consideration of the undesirable characteristics that light water slow neutron nuclear power plants have, in my judgment no new plants of this type should be constructed. The current fleet of 104 light water plants have around another 20 years of life expectancy and should be continued to be used, since they can provide environmentally benign electricity for the rest of their life. The long lived nuclear waste they produce can be used for 4th generation reactor fuel, thereby mitigating their waste disadvantage.

The new 4th generation fast neutron reactors do not have the disadvantages of the obsolete light water plants. In 2005 General Electric – Hitachi testified before congress that they can produce Prism 4th generation reactors for $1,300.00 per kWh. The cost of a coal fired electrical generation plant costs between $1000.00 - $1500.00 per kWh to construct. Comparing the two, a 4th generation nuclear reactor has about the same construction cost as an average coal plant. Besides the environmental and health problems produced by the use of coal, it also has to be continuously purchased to burn. The new 4th generation nuclear power plants uses radioactive waste for fuel and have no fuel cost.[85] Not only would these plants reduce and ultimately eliminate our stockpile of nuclear waste but it would also eliminate fossil fuel and uranium mining. Another great advantage of 4th generation nuclear is that it does not need an upgraded electric distribution grid. So as the old fossil fuel plants are decommissioned, a nuclear

---

[85]   *The 4th generation plants produce about 1 ton of radioactive waste per year for a 1 million kWh plant that requires 500 years to completely decay away.*

166

plant can simply take its place on the existing grid. The Prism power reactor is a standardized 311 mega watt modular design. By using a modular design the start up problems due to excessive scale up and non standardized plant design that occurred with the 2nd generation power plants in the 1960s – 1970s is avoided.

Well it's time to run the numbers to see how much it would cost to change to this type of power source. I will use the 300 million kWh value for the calculations which includes replacing coal, gas, and oil fired types of power plants. I will return later to how the peakers can be eliminated using base load reactors.[86]

~~~~~~~~~

The cost to replace fossil fuel generated electricity:

Total: $390.06 Billion.

Land Usage: 60,000 Acres.

Number of one million kW plants: 300 comprised of 947 modules.

---

[86]   *So we need 300 million kWh of electricity.*

Cost for reactors 300 million kWh X $1,300.00 / kWh = $390 Billion.

We need 300 one million kWh power plants comprised of 947 modules.

The average land requirements for a single plant is 200 acres. Thus, we need 300 plants X 200 acres = 60,000 acres.

The land cost is 60,000 acres X $1,000.00 / acre = $60,000,000.00

Total Cost $390.06 Billion

This will produce a 49% reduction in the US greenhouse gas output as well as other types of pollutants!

See below for a detailed breakdown.

~~~~~~~~~~~~~~

In order to get an idea about the status of development of fast neutron reactor technology a brief history is now appropriate. When commercial nuclear power was starting to be implemented it was known that the light water reactors would produce large quantities of long lived nuclear waste.

To address this problem the federal government implemented two approaches to deal with the waste products. In 1964 a fast neutron reactor named the EBR2 (experimental Breeder Reactor 2) was built to provide a means of developing a commercial fast neutron reactor design, the program was called PRIZM. In addition to having the objective of eliminating the impending nuclear waste problem, it was designed to have both a competitive construction and energy production cost relative to fossil fueled plants. The envisioned technology also had the goal of producing a process that would make its use impractical for production of bomb making materials. EBR2 / PRIZM was operated successfully until 1994 (30 years) and is the basis for the General Electric – Hitachi PRISM that is now available. It has resulted in a electrical generation reactor that has met all the goals originally sought. The second means of dealing with nuclear waste was to create a disposal site where radioactive waste could be buried and left to decay to a safe level. The current site where this may take place is Yucca Mountain. In order to fund the eventual storage costs after the site was developed at taxpayer expense and ready to accept waste materials, the

US government set up an escrow account in 1982. The escrow account was funded by the utilities who would pay into it at the rate of .1 cent per kWh of energy produced in their light water plants. The electric utilities have been unwilling to voluntarily pay for their clean up costs and have been fighting the .1 cent charge. Recently they have won in a DC court which ruled that these payments could be eliminated in 2014. The escrow account currently contains $30 billion. The estimated disposal costs for the nuclear waste they have produced is $175 billion, leaving a funding short fall of $145 Billion that would have to be payed by the taxpayer.[87]

How safe are the 4th generation nuclear power plants? One of the goals of the PRIZM project was to design an extremely safe reactor. This type of reactor does not have high pressure in the reactor vessel like the 2nd generation light water reactors that are now being used, where the water is pressurized to about 136 atmospheres. PRISM uses liquid sodium metal as a coolant and operates at ambient air pressure. The PRISM reactors have passive and active safety features: if the pumps used to circulate the liquid sodium failed, the coolant in the reactor would still circulate as a result of convection which works by gravity. A second safety feature would also come into play if coolant pump failure occurred, the reactor would leak out neutrons which are essential for the production of a chain reaction, automatically shutting the reactor down. By stopping the chain reaction, the production of heat is terminated eliminating the possibility of core melt down. A

---

[87]   *It should be noted that the eventual actual cost isn't really known since the Yucca Mountain facility isn't ready for operation so far $8 Billion has been spent on it. When it will be in operation isn't known at the present time.*

third safety system for shutting down the reactor employs control rods that would drop into the reactor core by gravity. A forth safety system is comprised of boron carbide balls that absorb the neutrons that cause the chain reaction.[liv] So as we can see the 4th generation reactors incorporate multiple redundant safety systems. Since half of them are passive there is no possibility of a core melt down as a result of operator error, as happened at Chernobyl and 3 Mile Island. Or have the reactor vessel rupture which occurred at Chernobyl as a result of excessive pressure. The only safety criticism of 4th generation reactors is the use of sodium as a coolant. Sodium metal reacts with water to form a caustic chemical (sodium hydroxide) similar to the lye used for cleaning stopped up drains. The Japanese have a fast neutron reactor (not of the same design as PRISM) where this type accident occurred and it did produce a significant chemical cleanup problem which required the plant to be shut down. This accident produced no problems with radioactivity or core meltdown. The problems with this plant are being corrected and it is anticipated that the plant will be restored to operation. The type of accident that happened at the Japanese plant isn't possible in the PRISM design.

In the PRISM design the heat is transferred from the reactor to the steam boiler through a secondary system. The secondary system still uses sodium, but the nuclear reactor itself is isolated from any potential contact with water.

Another advantage to the 4th generation type of plants are that they are comprised of standard modules (311,000 kW each), as already mentioned, they will not have the scale up problems that occurred with the start up of the 2nd generation plants. Since only one perfected design is being used it will also allow reductions in their cost if significant numbers are produced through the economies of large scale mass production.

While $390.06 billion is a lot of money to build these reactors, when this cost is considered in the light of how much the public is paying for the externalized costs of using coal ($345 billion per year from the Harvard study), 4th generation reactors are an extremely good investment since they would pay for themselves by eliminating the costs of coal in slightly less than 14 months. A potential source of funding that already exists may also be available. If you recall $30 billion is being held in escrow for the disposal of radioactive waste being stored at the light water plants. Fast neutron reactors are specifically designed to dispose of this waste. If the $30 billion were made available for construction, twenty three one million kilowatt 4th generation plants could be fabricated. Building these plants would Provide a "jump start' for the transition away from fossil fuels while reducing the stock pile of nuclear waste. The other nice aspect about using the $30 billion escrow would be that these first twenty three plants would essentially have no need to acquire further funding for their construction cost! Therefore, this would be an ideal way of starting the nuclear renaissance by using the escrow for the construction of publicly owned plants. At the present time 26% of the electricity generated in the United States is produced by publicly owned utilities. These publicly owned utilities have an average product cost 18% lower than privately owned utilities. I would also argue that public ownership in addition to producing much cheaper power would also have a much higher standard of management since their activities are a matter of public record. By having this type of transparency it would be difficult to cover up problems like the private power industry did. A publicly owned utility would also have the advantage of having no means of externalizing costs onto the public such as we saw for the Montana gold mine and more recently with the nuclear clean up escrow termination.[88]

By having public ownership the mechanism of transferring public resources to the owners of these businesses would be eliminated. Publicly owned institutions also have much greater access to state and federal resources (assuming these plants would be owned by municipalities or counties). Using the escrow money in this way would also provide a great incentive for start up since the cost of power to the public owners / subscribers would be extremely low, being comprised of only operating costs. Moreover, the lower costs of public ownership would also have a large impact on improving the prosperity of the people in the community. This would occur as a result of reduction of across the board business costs, as well as personal expenses. Another possibility would be to approach the federal government for plant construction funding to eliminate the 500 tonnes of bomb making nuclear waste. These plants would have to also have federal security forces and be modified to use this material as fuel. Using these materials are not a security danger, bomb making materials have to be diluted for fuel use, which makes this type of fuel useless for bombs.

In my view the best start up approach would be to present an educational program to correct the erroneous, obsolete, and PR disinformation that is creating the inaccurate prevailing impressions that most people have about nuclear power. This is essential because the local political establishment will not support this type of initiative without adequate public support. After adequate public support for

---

[88]   *Since the estimated net clean up costs without the use of 4th generation plants was $145 Billion this is 1,000 times greater than the public cost for the gold mine in Montana, WOW!*

construction of the new type of power plant is created, the conditions for county or municipal government officials to initiate construction would exist. The construction program could then quickly proceed by simply having a bond issue to raise the necessary funds. Once the plant is under construction or constructed then an attempt to gain access to the other (escrow and other) potential sources of funds could then be undertaken. If obtaining the additional funding is successful the acquired funds could be used to buy back the bonds or provide electric rate reductions.

If the suggested 300 plants are constructed they would reduce the amount of green-house gasses being produced by the United States by 32% for carbon dioxide, methane 11%, and nitrous oxide by 6%.

Since we are interested in reducing pollution and eliminating public costs derived from externalization of business expenses, another sector of our energy usage could be changed to utilize nuclear power. At the present time 30% of the fossil fuel energy used in the United States is for space heating.[89] It would be quite simple to replace the fossil fuel heating systems with electric heat, thereby eliminating the same plethora of fossil fuel expenses and environmental problems originating from this source too. At the present time by eliminating fossil fuels from this sector a further 30% reduction of emissions and the pollutants from mining would be realized.[90] According to

---

[89]   *The amount of energy usage in the US for space heating is 40% but 10% is provided by electric, giving us 40% - 10% = 30% from fossil fuels mostly from natural gas.*

[90]   *At the present time geothermal heating and air conditioning systems use 1/5 the amount of energy as conventional electric heating and cooling systems. If this technology were widely implemented a fewer number of power plants would be required in*

the EIA the amount of fossil fuel energy used for residential space heating was 6.42 quadrillion BTUs per year and for commercial space heating 3 5/8 quadrillion BTUs per year. So let's run the numbers to see what it would cost to replace fossil fuel heating with nuclear.[91]

~~~~~~~~~~

The total cost to replace fossil fuel space heating would be $460.69 Billion.

The number of modules required – 354.32 / .311 = 1139

The land required: 70,864 acres.

The number of reactor modules required: 1139

This would reduce US greenhouse gas emissions by 30%.

This would eliminate the energy equivalent of 1.43 billion barrels of oil per year!

~~~~~~~~~~

---

proportion to the level of geothermal usage. These systems are more costly than conventional heating systems though.

[91]   *We consume 10.55 quads of BTUs per year or 10.55 quads / (364.25 days/ year X 24 hr. / day) = 1.21 billion BTUs / hr.1 kWh = 3415 BTU per hr.*

So we need - 1.21 billion BTU / 3415 BTU/kWh = 354,319,180 kWh = 354.32 one million kWh plants.

The plant cost would be 354.32 million X $1,300 /kWh = $460.62 billion

Land cost 354.32 X 200 acres X $1000/acre = $ 70.9 million

Total cost is $460.62 billion + $70.9 million = $460.69 Billion

Railroads are another fossil fuel energy consumer that could be converted to electrical power derived from nuclear reactors. They currently consume 446,999,921 gallons of oil per year or 10,642,855 barrels of oil. If coal usage to fuel power plants were eliminated it would reduce railroad freight tonnage by 51%, thereby eliminating 51% of the oil consumed by trains giving a residue consumption of 5,215,000 barrels of oil. A barrel of oil when burnt produces 1,433 kg of carbon dioxide and if we do the arithmetic.[92] The amount of carbon dioxide eliminated by switching to nuclear power, thereby reducing train fuel usage would be:

~~~~~~~~~~~~

Carbon dioxide reduction = 7.473 million tonnes.

~~~~~~~~~~~~

Railroads are very efficient at moving freight, the amount of fuel used to move 4 tonnes of freight by rail will only move one tonne of freight by truck over the same distance. According to railroad industry analysis, if 10% of the long distance truck freight were switched to rail it would eliminate 23.8 million barrels of truck fuel usage. Once again doing the arithmetic:[93]

---

[92] *5,215,000 barrels of oil X 1,433 kg carbon dioxide per barrel / 1,000 kg per tonne = 7.473 million tonnes of carbon dioxide.*

[93] *23.8 million barrels of truck fuel X 1,433 kg carbon dioxide /barrel = 34.11 Million tons of carbon dioxide reduction per year.*

The number of power plants needed would be 5.8 billion kW / (364.25 days x 24 hr / day) = 6.6 million kW / hr.

Amount of carbon dioxide reduction: 34,105,400 tonnes.

The number of 1 million kWh plants needed: 6.6 comprised of 21 modules.

The cost is $8.625 Billion.

So by changing to an electric rail system and utilizing more inter-modal rail transport to offset some of the long distance truck freight a total reduction of 23.8 million barrels of oil has been achieved.

At the present time 492 million barrels of jet fuel are used per year, if half of the usage of air transport were switched to electric rail it would save a further 246 million barrels of oil. Let's run the numbers for this.[94]

The number of plants needed would be 6.6 and the cost would be 6.6 million X $1,300 = $8.625 Billion

The number of modules needed 6.6million / .311 = 21

[94]   So we need the energy equivalent of 246,000,000 barrels of oil.

The amount of carbon dioxide reduction: 246 million barrels of oil X 1,433 kg carbon dioxide / barrel = 353 million tonnes 246 million barrels of oil = 353 million tonnes.

~~~~~~~~~

Cost for nuclear power plants: $62.21 Billion.

Number of one million kW plants: 47.8 comprised of 154 modules

Amount of carbon dioxide averted: 353 million tonnes.

~~~~~~~~~

Summary for Nuclear Power

~~~~~~~~~

Number of 1 million kWh plants.............708.4 comprised of 2,278 nuclear reactor modules.

Cost ....$ 925.6 Billion [95]

Greenhouse Gas Reduction.....................86.2% (Note: The .2% came from switching urban mass transit to electric).

Nuclear Waste Elimination 708 Tonnes / year (42,480 tonnes over the 60 year life expectancy of these plants).

~~~~~~~~~

Let's consider the effects on cost through the economies of scale that would accrue since a large number of 4th generation modules would be required to achieve the reductions described above. To arrive at a more realistic rough cost we will assume a 10 year goal for completion of the project, which gives us a lot size of around 250 per

---

[95]   *The $910 Billion is based upon producing power plants in a lot size of one, i.e., as one offs, the actual cost if the lot sizes were larger say 300 modules would be much lower, so the above price is really high if we decide to convert to this energy source.*

year.[96] The unit cost adjustment is based upon my personal experience of estimating projects. Of course we were not making nuclear power plants and these values should only be considered an approximation for this type of product. When I worked in industry the cost for making larger lot sizes than one unit usually worked out to have the following cost reductions. If a lot of 4 units of the same thing were produced the cost would drop by around ¼. If a larger lot size of 10 items were produced the unit cost was about 2/3 of the cost of a one-off. These cost reductions would gradually continue down as the lot sizes became larger until they became quite large, whereupon, at some point no further cost reductions could be obtained. I will make an educated guess that the cost reductions described above probably are at least in the "ball park" of what could be expected. So if the $30 billion in the disposal escrow fund were used to place a blanket order we should get around 110 modules using the 2/3 discounted cost. Thus, we would get 34 one million kW plants comprised of 110 (311 megawatt) modules. If these funds were applied at the one-off price we would get 23 one million kW plants comprised of 74 (311 megawatt) modules. Since I have suggested that the new energy system be publicly owned, the fabrication of the electrical generation equipment could incorporate a contractual requirement that they would be produced domestically in the United States. By having this type of requirement large numbers of long term high paying domestic industrial jobs would be created. In addition to the jobs created by manufacturing the plants a large number of construction jobs would also be required to

---

[96] . *We will need to make around 2,500 modules assuming a .8% per year increase of electrical usage which is the current value. So 2,500 modules / 10 years = 250 modules / year.*

site them, and of course many additional high paying permanent jobs for the power plant operating, administrative, and maintenance personnel would be created.

One often hears the argument of why should we bother with this type of program since China is producing more carbon dioxide than the United States (in 2012 China produced 8.3 billion kilo-tonnes and the US produced 5.2 billion kilo-tonnes of carbon dioxide)? At the present time (2014) the United States has a debt of $2 trillion dollars with China, it seems likely that the Chinese would probably accept payment for part of this debt in nuclear power plants since they have a number of motivating circumstances. At the present time China has a severe energy shortage and pollution problem which they are having difficulties solving. Their remaining coal deposits are small and are dwindling very rapidly. Thus, the Chinese are facing an impending energy resource shortage. Of course other countries such as India are emerging as industrial giants and are also burning lots of coal, and experiencing coal energy related problems. These overseas problems could be mitigated if the United States could start up some type of program similar to the "Lend Lease" we had with the British during World War II. This type of approach would go a long way to solve their problems as well as the entire global pollution and energy situation. If this type of program were implemented the offshore nations using these reactors would need nuclear fuel for their new reactors, perhaps we could make some of our nuclear waste available for this purpose to offset the costs of the "Lend Lease" program. By using this approach we could eliminate the US stock pile of nuclear waste much more rapidly.

Agriculture is another area where significant reductions in excessive resource consumption could be easily achieved, as well as producing improvement of the environment. At the present time 9% of the United States greenhouse gas emissions originate from agricultural activities. The sources of these emissions are: 8% from carbon dioxide mostly from fossil fuel usage, 60 % from nitrous oxide, and the remaining 32 % from methane. The older conventional methods of vigorously tilling soil turns it over exposing it to the air. When soil is exposed to the air it produces carbon dioxide as a result of the oxidation of the buried organic components of the soil. These older methods also produce conditions favorable to soil erosion and depletion. According to Montgomery, soil oxidation from conventional plowing has contributed 1/3 of the excess carbon dioxide build up in the atmosphere since the 18th century.[97] In recent times new methods of tillage such as no till and conservation till have been developed that minimize or eliminate these problems. No till is the most effective of these techniques and is currently being used to cultivate 23% of the land in the United States. No till does have some limitations as to where it can be applied, being useful only in well drained soils. As a result of these improved tillage practices the soil in the United States now removes carbon dioxide from the air. The amount that is sequestered is small .4% but can be increased. Starting in the late 19th century the role of individual plant nutrients (fertilizers) were identified. These discoveries resulted in large scale applications of these substances to restore and maintain soil fertility. In the 20th century the use of these nutrients as well as pesticides and herbicides have been heavily promoted by the chemical industry as a panacea for every type of agricultural deficiency. This style of

---

[97] *. An input of atmospheric carbon dioxide requires up to 200 years to be removed from the air by natural processes.*

agriculture does have some drawbacks though, the application of nitrogen fertilizers particularly in excessive amounts causes the production of nitrous oxide in the soil. After this gas is produced in the soil it is then released into the atmosphere. Nitrous oxide is the largest source of agricultural greenhouse gas and persists for 120 years.[lv] The other significant source of agricultural nitrous oxide is from the deposition of manure and urine on fields. These deposits go through the same processes in the soil that convert nitrogen fertilizers into nitrous oxide. Methane which persists in the atmosphere for 12 years is the next highest agricultural greenhouse gas (32%). It originates primarily from livestock, rotting vegetation (7.2%), and burning crop residues (1%).

Bio-char a new/old technology offers a practical method of climatic remediation. Bio-char is a type of conditioned charcoal that is produced from agricultural and forestry waste. It has been estimated that it can permanently sequester 12% of the annual excess carbon from the atmosphere by incorporating it in the soil. It also reduces nitrous oxide emissions by 80% as well as all the methane, while improving agricultural productivity. Productivity is increased as a result of reduction of leaching of plant nutrients from the soil and improvement of water retention. In Brazil the Indians used bio-char as a soil amendment to enhance productivity of the low nutrient tropical soils that are typical of the amazon basin. The Indians utilized bio-char agriculture from 450 BC - 950AD. At the present time the bio-char treated areas of soil which are referred to as Terra Preta in the amazon basin is still productive and sought after by the local farmers for crop production or sale as compost. The estimated amount of Terra Preta in the Amazon is around 1% of its area or around twice the area of Great Britain. It has also been used in Ecuador, Peru, and Guyana in South America as well as in a few West

African locations. At the present time it is being very actively researched since it is not known how it will perform in all types of soil. It can, however, be safely deployed in the many types soils that have already been identified as appropriate.

In 2009, 13 Billion bushels of corn were grown requiring 35.1 million hectares (86.7 million acres), only 5 % of this crop is used for direct human consumption. Approximately 35.5 million head of cattle are slaughtered in the United States each year. Almost all of these animals are "finished" in cow CAFOs where they are fed mostly corn because it causes a very rapid gain in weight. On average 1273 kg (2800 lb) of corn is fed.[98] In order to produce the average CAFO cow it requires 284 gallons of oil (6.76 barrels), since cows normally graze on grass and not on corn we have managed to convert a source of food that is produced by solar energy to one that uses large quantities of oil (240 million barrels of oil per year). The amount of unnecessary carbon dioxide produced by the oil that is consumed using this method of agriculture is 76.1 million tonnes.

The amount of land that is required to produce all this corn for cattle feed is 4.8 million hectares (11.8 million acres). If the industrial CAFO style of beef production were abandoned and a return to grazing were used instead, it would free up 3 ½ % of the arable crop land. In order to obtain the maximum benefit from a return to grazing the expanded pasture land would need to be preconditioned with bio-char to reduce the greenhouse gas emissions that are produced by animal droppings and urine. Because the nutritional value per hectare of grazing lands have a

---

[98] *The amount of corn fed ranges from 1000 – 3500 lb per cow depending on its initial condition.*

significant amount of variation, I can't provide a realistic estimate of the amount of animals that could be produced on the land changed from crops to conventional pasture. So what I shall do is to fudge here by substituting types of livestock that utilize feed more efficiently than cattle such as poultry.[99] Since what is being suggested is a reestablishment of native prairie grasses no further nitrogen would be added through the use of mineral fertilizers or legumes that also produce nitrous oxide in managed pastures. The growth of the prairie plants would also directly sequester about 2.6 million tonnes of carbon from the atmosphere per year.[lvi] By instituting grazing and eliminating CAFO type of live stock production and its associated cattle manure storage impoundments, a further reduction of around 22 million tonnes of methane and 9 million tonnes of nitrous oxide per year could be achieved. Moreover, if the marginal agricultural lands that are currently being used to produce live-stock feed crops through the use of irrigation were converted to grasslands dramatic reductions of water usage could be achieved.[100]

---

[99] *. The amount of feed needed for to produce one kilogram of animal weight gain is as follows: cattle 7– 10 kg feed; swine 3- 4 kg feed; poultry 2 kg feed.*

[100] *Water shortages in California and the south western states that are highly dependent on irrigation is becoming a big problem. 5.4 cubic kilometers of water is diverted from the Colorado river to California each year. By building enough 4th generation base load nuclear plants to fulfill the intermittent high consumption periods, the excess power they produce during times of low demand could be used to operate desalination plants. The fresh water from these plants would greatly diminish the water demands placed on the Colorado river, and also provide greater water security for California. By having greater amounts of Colorado river water available for use by the south western states, the amount of well*

The second largest source of live-stock greenhouse gasses originates from swine CAFOs which produce around 19 ½ million tonnes of methane and about 2 million tonnes of nitrous oxide. These emissions could also be mitigated by implementation of the same types of managed live-stock waste as described above. By eliminating swine CAFOs the waste production becomes dispersed enough so that a daily spread of the waste on fields can be used. A daily manure spread greatly reduces greenhouse gas production since only tiny amounts of methane and no nitrous oxide are produced. The remaining types of livestock; horses, poultry, sheep, and goats produce trivial amounts of greenhouse gasses and will not be given further consideration. If the remaining old style type of soil tilling were eliminated, a further 2.1 % reduction of carbon emissions could be obtained by replacement with organic no-till and conservation tillage.[101]

A highly efficient form of grazing is practiced at Polyface farms. This system is designed to mimic natural prairie processes. In order to achieve this goal it incorporates a succession of live-stock types combined with intensive rotational grazing on perennial poly-culture pastures.[lvii]

---

water usage could be diminished. At the present time the Ogallala aquifer where much of this well water originates has fallen as much as 300 feet since 1940. Some estimates indicate that this aquifer could be essentially exhausted by 2028. A 2013 study by Leonard F. Konikou of the USGS indicated that the amount of water drained from this aquifer between 2001-2008 was as great as 32% of the entire water withdrawal of the 20th century!

[101] According to Purdue University's figures using no-till also reduces the amount of labor used to plant a five hundred acre farm by 225 hours. Moreover, it also greatly reduces oil consumption because of the reduced amount machine operation time.

The livestock types that are used are cows, chickens, turkeys, rabbits, and pigs.

Let's take a brief and simplified look at how the system developed by Salatin works. He utilizes small pastures of slightly less than a hectare (2 acres) comprised of plants that are not quite ready to start producing seed.[102] He creates a paddock using portable electric fences and allows the cows to enter for one day only.[103] Next a new pasture is created by simply moving the electric fence. The animals move into the newly created area by themselves since they have learned that is where their next meal is. Of course they leave many droppings behind in the grazed pasture which attracts flies. The flies lay eggs in the cow pies that matures into full grown maggots after about 1 1/2 weeks. Next chickens who reside in *egg-mobiles* or *feathernet* units a (type of mobile "chicken coop") is moved into the old pasture when the plants are between 2 1/2 – 7 1/2 cm (1 – 3") high for grazing. Maggots are a preferred food by chickens and also act to reduce the amount of grain fed. The combination of these pasture foods reduce the amount of grain usage by 30 % during the foraging season.[lviii] In order for the chickens to obtain these delicacies they scratch the cow pats with their feet in order to expose the fly larva, thereby distributing the manure (fertilizer) over the pasture while obtaining free food. This also produces the added benefit of keeping the fly population at a reduced level.

How does this system compare to the CAFO method? Before we conduct our comparison one should be aware

---

[102] The pastures are comprised of fescue, timothy, red & white clover, plantain, orchard grass, dandelion, chicory, and wild carrots (p. 21).

[103] The quality of the forage is very high as a result of excellent soil fertility which allows the high number of animals to use the paddock with out overgrazing it.

that the Polyface farm was established by William and Lucille Salatin in 1961 on a severely depleted farm in the Shenandoah valley located near Staunton, Virginia where a continuing and successful effort has been made to upgrade the soil using organic methods. According to Joel Salatin this farm is 5 times more productive than the conventionally operated farms in the same area.[104] After converting the *cow days/acre/ year* he uses as a measurement to the more familiar acres/cow, we find that it requires .89 ha (2.2 acres) to produce a steer.[105] Now let's consider how much it requires to produce a steer by industrial CAFO's. I will use the 80 cow days/year that is the average for the Staunton area and the USDA values for CAFO finishing, i.e., 140 days in the CAFO where 1,270 kg (2,800 lb.) of feed is used.[106] Referring to the

---

[104]   He uses a measurement called a cow-day which is the amount a cow eats in one day. The poly face value is 400 cow-days/year versus 80 cow-days/year for conventional farms.

[105]   I am assuming that grass fed steers are slaughtered at 24 months (from answers.yahoo.com) , and a CAFO steer is 18 months old (USDA). So we have 400 cow-days per year per acre/ 365 days per year = 1.1 acres a year. And since they require 2 years to grow it requires = 2.2 acres or .89 ha.

[106]   To find number of bushels of corn; 2800#/56#/bushel = 50 bu.

Corn yield = 140 bu/ acre and 50 bu/140 bu/acre = .36 acres needed for grain fed at feed lot.

Converting cow day years: 365 days/year / 80 cow day/ year =4.25625 acres for a year.

Since a steer is slaughtered at 1.5 years it remains at the starter farm (assuming it's pastured) for 1.5 years X 365 days/year – 140 days = 408 days or 408 days /365 days/year = 1.118 years

Thus we find that our steer requires 4.25625 acres/year X 1.118 years = 4.758 acres for pasture + .36 acres for CAFO feed = 5.118 acres per steer.

calculation in footnote 105 we find that 2.07 ha (5.118 acres) are needed to feed a industrial steer compared to .89 ha (2.2 acres) using the Polyface method which is 233% more efficient! Moreover, this number doesn't include the reduced amount of feed required by the chickens during the foraging season which further increases this level of efficiency.

Another great benefit of using this system is that it rapidly improves and augments the depleted and eroded soils. To illustrate this point, when Salatin purchased their farm it had large barren areas where the entire layer of top soil had been eroded away leaving the bed rock exposed. By using the method of intensive rotational grazing and on site composting they were able to correct the fertility deficiencies on this land. In 2010 about 50 years after the project was started even the completely barren rock patches had accumulated at least 20 cm (8") of top soil. This is a remarkable achievement considering that under natural conditions it requires 500 years to produce 25 mm (1") of top soil.

Let's take a deeper look at how this process works. If we consider how perennial pasture plants grow, their fastest growth takes place during the period just before the plant engages in seed production. As these plants grow both the top of the plant and the roots increase in size. When they are grazed the top of the plant is removed which produces a plant with roots that are too big for the size of the top. The plant responds to this imbalance by shrinking the root size by letting portions of them die. Thus, if the plants are grazed at the appropriate time in their life cycle huge amounts of organic material is added to the soil which improves it's fertility and depth.

Taking a closer look we find that the composition of the root material being added it is about 60% carbon[lix], thus by using this process not only is the soil fertility being improved but large amounts of carbon is being sequestered. A further advantage of this system is that ruminants such as cattle that are grazed on poly-culture pastures their digestive systems produce less methane from enteric fermentation.

Now let's make an educated guess about how much carbon sequestration could be expected from a change to the Salatin method. A CAFO steer consumes the amount of grain produced by .15 ha (.36 acre) and ~34 million steers are slaughtered each year, it requires 5.1 million ha (12.6 million acres) to produce this feed. I will use a conservative estimate of 10 cm (4") of humus formation 1/2 of the value published for Polyface farm (see foot note for calculations).[107] The amount of carbon that this system

---

[107]   According to the Encyclopedia Britannica humus is 60% carbon, and the USGS gives a value of 797 – 1055 kg / cubic meter. I will use the mean value of 926 kg / m³.

so each cubic meter of humus would cover: 10cm of humus / 100 cm/m = 10 square meters. And there are 10,000 sq. meters in a ha. Thus the amount of humus produced in 50 years would be 10,000 m² / 10 m² per cubic meter of humus = 1,000 cubic meters/ ha. And in 1 year we find that it is 1,000 cubic meters humus /50 years = 20 cubic meters per year.

And 1 cubic meter of humus contains 926 kg X .6 carbon = 555.6 kg carbon.

Thus 20 cubic meters humus / ha / year X 555.6 kg = 1111.2 kg / ha / year or 11.112 tonnes / ha / year.

And multiplying our 5.1 million ha X 11.112 tonnes / ha = 56.67 million tonnes of carbon per year. Which would produce 207.8 million tonnes of carbon dioxide (the weight increases because you are adding oxygen to the carbon to produce carbon dioxide).

would store in the soil is 56.6 million metric tons of carbon, thereby, averting the formation of 207.8 million metric tons of carbon dioxide each year (1 kg of carbon produces 3.67 kg of carbon dioxide). This value is certainly only a fraction of the carbon this system could actually sequester since it does not include the additional grain reductions produced by its combination with poultry. Pigs can also be grazed using the Salatin method or a similar system developed by the Rodale Institute which would produce further sequestration of great amounts of carbon (see CAFO debunker No. 5 in the appendix for further information). Note: I am not going to include these values in the totals below since not enough information is available about how widely this system can be implemented. It certainly could be used wherever row crops such as corn is currently planted as described above.

Since the implementation of these agricultural practices would create significant reductions in the amount of land needed to produce livestock based food, this land could be turned back into natural areas. Edward O. Wilson estimates that if 50 % of the worlds land area was conserved in wild life corridors connecting biological hot spots, the loss of species could be reduced from an estimated 50% by the end of the 21st century in the business as usual scenario to a 15% loss.[ix]

This system does have a few disadvantages though. Since it is not heavily subsidized the retail cost of its food is higher. It also doesn't lend itself to large scale industrial production which relies on a highly uniform production system, i.e., monocrops, CAFO style facilities, and huge centralized processing plants.

From a food consumers view point you do get a superior product which is lower in fat, better balanced and with higher levels of nutrients (I personally think the flavor is better too). When one considers consuming this type of food I think the old adage, you get what you pay for is right on the mark.

The question is what can we do as individuals to change to these kinds of systems? It's really quite simple! Vote with your dollars and buy these kinds of environmentally benign products. If you are interested in using this kind of food a few tips about what to look for and where to find these kinds of products is included in the appendix titled "Food Tips: Or how to avoid ersatz foodish products."

Before proceeding further lets total up the amount of the agricultural greenhouse gas emissions that have been averted and sequestered. Since some uncertainty exists about the amount of land that can be treated with bio-char, I will make the assumption that 80% can be treated (10% carbon sequestration instead of 12%)

Origin	Amount	% Reduction
Bio-char	652.6 billion tonnes	10
Cattle & Swine	126.4 million tonnes	00194
Improved Tillage	137 million tonnes	0021
Total	**656.4 billion tonnes**	**10.04**

So far all of these agricultural carbon reductions and offsets have been accomplished simply by implementing improved methods, existing new technologies, rearrangement of land usage practices.

Another possible source of greenhouse gas reduction could be achieved by minor dietary changes from one meat source to another or to dairy products or eggs.[108] If a shift away from the least feed efficient livestock sources of meat (cattle and swine) to the most efficient source poultry, some further reductions in crop land usage and greenhouse gas production could be achieved. Let's consider what a 10% shift from inefficient cattle and swine to poultry would accomplish (since we have performed the same types of calculations above I will skip the arithmetic and just provide the values).

~~~~~~~~~~

Net Tonnage of CO2 equivalence (From enteric fermentation).

Cattle...............13.37 million tonnes.

Swine................5.87 million tonnes.

Percentage of greenhouse gas reduction, .4%

Note: the above figures assume that the suggested implementation of Bio-char amendments and that improved agricultural practices have taken place.

~~~~~~~~~~

Let's consider what the effects a 10% reduction of meat consumption would be. In this case CAFO oil cows will be used for the example (note: swine would produce about ½ of the reductions from cattle, poultry about ¼. We would

---

[108] *Dairy and eggs are an even more efficient than any of the meat or poultry food sources. They require about 1/5 the amount of animal food to produce the same amount of nutritional value as does beef.*

eat 3,550,000 fewer cows per year and reduce our oil consumption by 24,000,000 barrels. This would produce 7.61 billion tonnes of carbon dioxide reduction per year, and take 478,871 ha (1.2 million acres) of crop land out of production. If this land were reforested the trees would sequester an additional 325,720 tonnes of carbon dioxide each year. An additional 13.8 million tonnes of carbon dioxide equivalents would be averted that originate from emissions caused by enteric fermentation, managed livestock waste, and pasture[109]. The total for a 10% reduction of beef consumption is:

~~~~O~~~~

Total : 7.624 Billion tonnes of carbon dioxide reduction.

---

[109] A shift from CAFO beef to grazed dairy is another possible area where large reductions of green house gasses could be achieved. Dairy produces five times the amount of food calories for each hectare as is produced by beef. This increase in efficiency would greatly reduce the amount of cattle needed to produce a given amount of food, thereby producing a corresponding reduction in the amount of land land used to feed them. The composition of pasture plants also affects the amount of methane produced by cattle. This is important because cattle grazed on poly-culture pastures instead of cold weather monocrops such as alfalfa produces lower amounts of methane from enteric fermentation. Poly-culture pastures also have the further advantage of producing higher quality food. A further advantage of these types of pastures is that they are also much more resistant to environmental stresses, such as periods of low precipitation, predation from pests and disease.

A 1.2% reduction in the annual US greenhouse gas output.

~~~~~0~~~~~

The actual value just presented is low since the early part of the life of CAFO oil cattle takes place on pasture some of which is improved by tilling and the addition of fertilizers that produce greenhouse gas emissions. By taking this land out of production and being reforested, additional carbon offsets would accrue which hasn't been added in.

One of my friends Roc is a recently retired professor who taught classes in nutrition. For robust health he recommends a low meat diet and considers poultry to be the best type of meat to consume.[lxi] If we look at the recommendations of several religions, meat isn't considered to be a preferred source of food. For example some Christian sects refrain from eating meat one day a week. If one looks in the Christian Bible Adam and Eve did not consume meat. Meat only arises as a topic much later in the 8th chapter, Romans, where God doesn't view meat consumption favorably but allows it because people are weak. The Benedictine monks eat only slight amounts of meat and St Francis of Assisi never ate meat at all. The Buddhists go further and recommend a vegetarian diet. Obviously high consumption of meat isn't necessary to maintain good health and one needn't be concerned about reducing this part of the diet out of fear of acquiring a nutrient deficiency. Small dietary changes however, can have large impacts and I would urge you to consider your food choices in light of the above discussion.

Forestry: According to the EPA in 2012 forests were providing a 15% carbon offset (absorbing 15% of our carbon dioxide emissions). Looking at the table on page 68 of the 2008 U.S. Agriculture and Forestry Greenhouse Gas

Inventory we can derive the average amount of carbon sequestered per hectare as .628 tonnes each year. Forests are an effective means of sequestering greenhouse gasses. Thus, by increasing the amount of forest cover proportional reductions could be accrued. Some places where significant forest expansion could be accomplished are; replanting areas that have been cut for timber, planting alongside roads and streams, and expansion of the shelter belt[110] which currently occupies only 22 million ha.

At the present time the impairment of forest growth as a result of acid rain significantly diminishes the amount of carbon being sequestered by the sick trees. Once the shift from fossil fuels to nuclear power takes place the elimination of acid rain will allow these plants to regain their vigor and increase the amount of carbon sequestered in these areas.

At this point let's total up the amount of greenhouse gas reductions:

Nuclear power.......................86.2 %

Agriculture............................11.64 %

Total:....................................97.84%

Note: The amounts from transportation mentioned earlier and dietary shifts are not included in the above total.

---

[110]   . *Forested land can be used for grazing, so an expansion of forests would provide a large carbon dioxide offset and at the same time additional grazing area.*

The remaining large contributors of greenhouse gas emissions are from transportation 28% and industrial consumption 20%. Transportation is an area where significant reductions have taken place with more reductions anticipated in the near future. When I looked into the various energy technologies about 10 years ago the United States consumed 20.4 million barrels of oil each day for transportation. Since then oil consumption has dropped to 18.9 million barrels per day, a 7.4 % decline in oil consumption which reduced our national carbon dioxide output by about 2.1%. The recent updated mileage standards for cars have a target of 54.5 mpg by 2025. The increase automotive mileage should produce an anticipated reduction of about 2 million barrels per day about 10.6% of our current consumption. This reduction of oil usage will produce a further 3% decrease of the national carbon dioxide output. These projections are based upon the assumption that the car usage patterns will remain similar to the present which may not be the case (we will return to his later).

Let's consider some applications of electric power for transportation. When I was a child living in Chicago the mass transit system was powered by electricity. It was comprised of the "L" which is light passenger rail and is still in operation. The city also operated electrically powered buses (Plate. 1) and street cars popularly referred to as green hornets. The green hornets were another form of light rail that utilized tracks embedded in the streets. The electricity to power the buses and Green Hornets were picked up from overhead electric wires which can be clearly seen above the bus in Plate 1. Shortly before we moved from Chicago the Green Hornets were eliminated and the buses changed to diesel. The decline in air quality was very obvious, the air really stunk from diesel. I recall asking my father why we changed from the nice quiet clean electrically powered transport to diesel? He said that the diesel buses didn't need to stay by the wires. Of course looking back on this now his explanation wasn't correct because public transport always follows the same route. Some years later it turned out that the automotive industry hatched a plan to eliminate electrically powered public transportation. The method that was used was to acquire a controlling interest in these transportation systems and put them out of business. This was accomplished by setting up fronts to conceal their maneuvers to gain control of electrically powered public transport systems. Once the electrically powered systems were gone, the public need for mass transportation created the conditions that enabled these companies to replace the electrical system with their inferior diesel powered buses. This greatly increased sales of their products and was also found by the courts to be illegal.[111]

Recently one of my friends took a trip to Mexico and he noticed that they were operating electric buses. He mentioned to the driver that these same types of buses had been used in Chicago. The driver replied that they were the very same buses we had. They had been purchased from the Chicago Transit Authority years ago. These seem to have been good buses they were purchased used and still in operation about 60 years later. I wonder if any of the original diesel buses are still in operation and still polluting the air? I suspect not.

---

[111] *Another triumph of corporate greed over public welfare. Many years after slowly wending it's way through the courts when the litigation had become moot as far as electrical transport was concerned they were fined $1.00.*

Plate 1. Electric Bus Courtesy of Kevin Zolkiewicz.

The point of this reminiscence is that it is a description of a well tested practical method of urban electrical transportation. According to the information published by the United States Department of Transportation in 2011, diesel powered buses consumed 4.6 million barrels of diesel oil a shift back to electrically power buses would eliminate the use of all this oil, and if supplied by pollution free electrical nuclear power it would reduce our carbon dioxide output by 6.6 million tonnes per year. The conversion to electrically powered urban transportation would also reduce large numbers of illnesses originating from the diesel exhaust particles, thereby producing large reductions in health care costs and improved quality of life. [112] If the suggested billion tonnes of long distance truck

---

[112]   *It is estimated that if filters that eliminated diesel exhaust particles larger than 2.5 microns, a $100 billion reduction in health care costs would be realized. The cost to outfit the U.S. Diesel fleet with filters would be $10 billion. The diesel powered transportation*

freight could be shifted to inter-model rail, the following benefits would occur; consumption of 29.8 million barrels of oil and the production of 42.7 million tonnes of carbon dioxide.

Next I would like to consider a possible expansion of the electrical urban mass-transit system to also carry freight. Before we start though, the following proposal would require the development of electrically powered trucks. Therefore, the following presentation is being included to provide planning guidelines for an urban mass-transit system that would have the capability for the expansion of power line capacity to accommodate the anticipated higher demand. It is also being included to provide an example of a simple achievable cost effective type of technical innovation that could quickly have a large impact on our environmental problems.

After deducting the one billion tonnes of long distance freight we still have another 8 billion tonnes of freight that is transported shorter distances. Most of this short distance freight is moved around in urban areas where 80.7 % of the U.S. Population resides. It seems that it would be feasible to transport urban freight using electrically powered trucks. These trucks could use overhead power lines for most of their journey to get them close to their destination. The remaining short distance to their delivery location could be accomplished by battery. My recollection of traveling using the public transit system in Chicago was that we seldom had to walk more than ¾ km (½ mile) to our destination. So a battery that could power a truck for maybe 3 km (2 miles) or less would be required to have a

---

industry has vigorously resisted the use of these filters and prevented requirements for air quality improvements in this area. Once again corporate greed triumphs over public benefits.

practical electric urban freight system. The electrically powered trucks could simply be created by incorporating already existing electric traction and control systems in a truck chassis and body, with the addition of the new transportation batteries all of which currently exist.

Let's consider how this could effect fossil fuel consumption. Since 80.7 % of the population lives in urban areas I will make the assumption that these people consume 80.7 % of the remaining 8 billion tonnes of truck freight and that 75 % of this freight is within the delivery range of the electrically powered trucks, which works out to 5.22 billion tonnes of freight. According to the U.S. Department of Transportation in 2011 trucks used 585.2 million barrels of oil to move all 9 billion tonnes of freight, which also produced 839 million tonnes of carbon dioxide. To move 75% of the urban freight by nuclear electric 254.6 million barrels of oil which produces 365 million tonnes of carbon dioxide would be eliminated. This would provide a further green-house gas reduction of 6.8 % and a 3.7% reduction in our oil consumption. Note: since the freight system requires development it is not being included in the cost, energy balance, and greenhouse gas tallies, but can be kept in mind as a likely near future technical panacea.

To sum up, the reductions of fossil fuel usage if urban diesel buses are discontinued and a return to the electrical bus system is implemented it would result in an additional reduction of 4.6 million barrels of annual oil usage and 6.6 million tonnes of carbon dioxide. In order to produce the electrical power for the entire transportation proposal including trucks, an additional 178.5 one million kWh plants containing 574 modules would be needed. The cost would be $232 billion at the one off price, or $155.44 billion at the discounted high volume rate.

After this point I will no longer be able to produce further quantifiable estimates, so let's total up what has been accomplished.

~~~~~~~~~~

Sector.....................................% Carbon Dioxide Reduction

Agriculture..............................11.64

Nuclear Power..........................86.2

Transportation..........................3.73 (2.52 million barrels / day = 13% reduction)

Total..101.57%

~~~~~~~~~~~

Batteries have four significant deficiencies for use in battery powered vehicles. The first shortcoming is that batteries for their size don't store a lot of energy and as a result of their low storage capacity, they are relatively heavy. The second shortcoming is that the amount of power that they can be tapped for is dependent on temperature. At lower temperatures the amount of available energy is less.[113]

A third problem is the amount of time that is required to recharge them. The last major problem is that large storage capacity batteries are expensive. If we consider the above discussion about the use of overhead electric power lines some of the problems just described can be reduced. The

---

[113] . *The degree of temperature sensitivity varies according to the type of battery.*

batteries carried in freight or passenger vehicles could be undergoing recharge while the vehicle is being used when it is tapping energy from overhead lines as suggested for urban areas. By recharging while traveling, alleviation of recharge down time would occur. An insulating blanket or simple battery heater powered from the overhead energy system could be incorporated in vehicles used in colder climates, thereby reducing the temperature sensitivity problem. Another approach to address the range problem would be to design the vehicles for standardized demountable batteries. By having the capability for rapid battery changes a great increase of vehicle range and/or a reduction in battery size could be achieved. Another advantage of a demountable battery system would be for use in long distance commuting or routine delivery service where inadequate time for recharging exists. A depleted battery simply could be left at the destination where it could be recharged and exchanged for a fresh battery. By using this system a doubling of the vehicles battery powered range or a reduction in battery size could be realized. It would also be simple to establish battery service stations where one could simply exchange a dead battery for a fresh one for a fee. By having a network of battery service stations a great increase of the utility of these vehicles could be produced, roughly on par with petroleum powered car transportation. At the present time a few rechargeable battery powered electric commuter cars are available and they suffer from the deficiencies described above. The battery powered cars that are currently available have ranges from around 98 km - 438 km (60 – 267 miles). The "gas mileage" they get is around the equivalent of 48 km per liter (110 miles per gallon) and require 4 to 10 hours to charge. Their prices are generally mid-range. Probably the greatest deterrent to their more widespread use is their charging time. By adopting the overhead power and demountable battery system just described their battery recharging deficiencies could be greatly reduced without sacrificing or possibly even increasing their range. If the batteries for transportation become standardized one could also expect a reduction of their cost because of the greater scale of production. A further advantage to using this type of system is that the cost of nuclear generated power is lower than other sources. By having lower transportation energy costs, further reductions in freight costs would be expected as well as lower personal transportation costs. These cost reductions are additive which would significantly benefit the consumer. Using inexpensive nuclear electricity would in effect reduce the cost for the electric transportation by around 20 %, and be about 5 - 6 times cheaper to operate than fossil fueled cars and trucks, wow! A final advantage is that these systems are more durable and reliable. They have much lower maintenance costs than internal combustion powered transportation, no oil changes, filters, fan belts, radiator hoses, PCV valves, etc.

According to published US Government information one of the primary objectives for obtaining the 54.5 mpg car mileage goal is to reduce the need to import oil by 2 million barrels/day. At the present time the United States is importing 18.9 million barrels of oil each day. By switching to the suggested electrically powered transportation system our oil consumption would fall by 2.52 million barrels per day exceeding the needed reduction of 2 million barrels per day.

Achieving the goal of eliminating oil imports is exceedingly important. At the present time a great expansion of fossil fueled transportation is occurring throughout the world. For example, China is producing and adding 14,000 new cars to their existing fleet every day. The result is that the amount of oil being consumed is rapidly increasing, placing progressively greater demands on a declining non-renewable resource. The result of this situation is that ever increasing levels of competition for this resource is taking place between nations, creating tensions that has and will produce wars.[114] Another problem with oil imports are that most of it originates from the middle east where a highly volatile political situation prevails. This fluid political atmosphere greatly compounds the already rapidly increasing geopolitical instability caused by oil's declining availability. By eliminating imported oil a reduction in international tensions would result. As a result of reduced geopolitical tensions and need for oil it would no longer be necessary to ensure the continued flow of existing foreign oil supplies. Thus, by eliminating middle eastern oil imports no rational would exist to maintain a military presence in the middle east. Thereby, providing the possibility of a large reduction in the size of the U.S. Military and the associated military expenditures.

Everything that has been discussed so far with the exception of a few minor things[115] can be quickly and

---

[114]   . *If you consider the two recent wars the United States has had with Iraq there can be no doubt that securing a reliable source of imported oil was one of the primary factors. In fact going back to the first Gulf war when Iraq invaded Kuwait, President Bush 1 explicitly stated that maintaining US oil supply was vital to our national interest and threats to this supply would not be tolerated.*

[115]   *The minor things are establishing a national transportation standard for standard battery configurations. This would make a practical demountable battery system to be established enabling the formation of a battery service station network. With this network the disadvantages of battery charging time hindering long range battery powered travel would be eliminated.*

easily accomplished, because they already exist, are adequate for the task, are cost effective, and can be immediately implemented. The result would be a reduction of the price of everything, improved health, and a huge reduction in the causes of the environmental problems that currently prevail. In order to accomplish this task approximately 2,500 modules would be required and if the rate of construction was at the suggested 300 module lot size per year it would take about 8 1/3 years to accomplish. Since at this scale of production significant reductions in costs would be achieved, I will make an educated guess based upon my experience, that an approximate cost for the project would be around $677 billion at the discounted rate. So if we add the $345 billion in savings estimated by the Harvard study for elimination of coal and the additional $80 billion in reduced health care costs from reducing diesel particulates, we could expect this program to pay for it self in 1 3/4 years.

The discussion of the challenges and some possible solutions that has been presented in this text has largely been focused on the United States. While the United States has been a major contributor to the current level of pollution, the environmental problems that we must overcome are global in nature. Therefore, an international cooperative effort will be needed to resolve it. In order to engender the necessary true cooperative spirit required to effect the profound changes that must be made, the extractive mode of relationships that currently prevail will have to be attenuated. In the final part of this guide we will explore some possible modifications in our political and social system that can produce the desired results and also provide an improvement in our general life situation.

Structural Social Change- In order to effect the desired changes, in this section we will consider the following:

1. Characterize the likely conditions arising from current trends that will form the environmental matrix we will have to cope with,

2. provide a few suggestions that will reduce the currently prevailing rigidity in the U.S. Political system,

3. propose a reorganization of the basic (grass roots) social pattern to provide an appropriate system able to operate effectively in the new environment.

The environmental factors that will probably exist are; a climatic regime that is expected to have greater variability, while the shorter term weather patterns are expected to incorporate larger but more persistent swings. The percentage of strong storms and other forms of extreme weather is expected to increase. The effects of climatic change will also produce a redistribution of the current moisture patterns. These changes in the weather regimes are expected to produce a reduction in agricultural output. Sea level rise is expected to encroach upon coastal areas that are generally heavily populated and developed.[116] Diminishing amounts of resources, particularly fossil fuels will be available. The greater amplitude of weather

---

[116] *Recently the Florida legislature past a resolution to divide the state in half. This occurred as a result of sea level rise impacting the southern more populated area which averages only 50 feet above sea level (the northern half enjoys much higher elevations). Because the remediation has been very expensive and mostly paid for by the southern inhabitants. They feel that the state hasn't been fair in the apportionment of tax derived funds to offset these expenses. Thus, we are starting to see some of the political ramifications of climate change.*

variation will produce increasing costs for maintenance of infrastructure.

At the present time we are relying on a global economy to provide many of the goods and services that support us. This type of economy relies upon opportunistic exploitation of locally favorable conditions and economies of scale. The types of conditions that are typically sought out are; areas with easily trappable resources, governments that allow or can be induced to accept externalization of business expenses and give concessions, ineffectively or unregulated labor practices, low labor costs, poor or nonexistent environmental regulations, and places where the populations can be made dependent upon the presence of the remotely owned multinational corporation. Economies of scale are achieved by consolidation of production and specialization. Basically with this type of system you have all your eggs in a few baskets where systemic interruptions, perturbations or breakdowns can produce significant disruptions or collapse. An example of these characteristics can be seen in the manufacture of computer key boards which are currently being made in only two locations, China and Taiwan.

The general trajectory of the pattern of these international enterprises mode of operation is as follows. In most cases these enterprises take root in a new location by offering somewhat higher wages than locally prevails, usually some where in the third world. The higher wages attracts employees often with experience or skills relevant to the activities of local enterprises, thereby depriving them of trained personnel. The international presence also establishes the conditions where the lower cost merchandise from the global economy becomes available which the new wage earners partake of. The result of these introduced changes are that many of the locally produced items are supplanted and the local economies become simplified. Because of the reduced production diversity the host community loses its organic capacity to produce necessities. As time progresses the constellation of desirable conditions that originally attracted the international company will change. These changes typically provide the impetus for the company to seek and find some new more desirable location where it will relocate their operation. After the company leaves its economic input is removed from the community where it was located and the local economy collapses. Generally these communities experience great difficulty regaining their original vitality since the capability for local production has ceased to exist. In general the area becomes impoverished and out of necessity many of the inhabitants are forced to migrate in the hope of finding an adequate life situation.

Not only does this style of economic activity simplify and cause eventual hard ships in the third world communities that are exploited but also diminishes the organic capabilities of the first world nations. In the 1990's these multinational enterprises launched a vigorous campaign to induce the politicians to establish a legal framework to support their practices. The desired legal framework they were promoting was embodied in the World Trade Organization (WTO) treaty. This international treaty was designed to promote globalization and was instituted in the 1990's. Let's examine the effects this has had in the industrialized nations. If the United States is considered, since the implementation of the WTO over half of the nations productive capacity has been lost. What has emerged is a simplified type of dependent economy. The economic simplification has produced a reduced local and/or regional manufacturing capability with the ability to produce a full array of necessities. For example, the number of fasteners that are domestically produced is minute. Very few of the most commonly used items such as clothing, shoes, electronics, etc., have any significant level of domestic production. The point is that the highly dispersed specialized nature of the globalized means of production is completely dependent upon having a continuation of the favorable conditions it's based upon. It is really quite fragile. If the sorts of environmental changes that are being predicted take place accompanied by scarce resource availability, the globalized means of production will not remain viable. Moreover, the condition of scarcity or absence of necessities often produce social, political, and economic disintegration as described earlier since they are an interlocked system.

The question is how can the greed based economic system that is the wellspring of our environmental and social difficulties be phased out without producing undesirable dislocations? Since this system does not exist in a vacuum there is much that can be done in the matrix of the individual and collective spheres. Considering this matrix more closely it becomes apparent that it requires a social system that tolerates and fosters an entire supporting suite of afflictive states of mind. So in order to change to a society that doesn't embody greed as the prime mover, the current social system needs to be overhauled. This can be accomplished easily and painlessly by incorporating psychological antidotes to attenuate individual greed and systemic social mechanisms that can not effectively embody its activity.

The place where effective changes can be made is in the individual character of people. Character is simply the way a suite traits are constellated. These traits are selected from an inherent inventory possessed by all persons. The degree of presence in the personality's constellation of a particular trait is dependent upon its frequency of appearance and associated level of intensity. Increased frequency augments its presence as well as the level of intensity. Both are governed by the conditions that gave rise to its manifestation. Much of the operation of these traits in our daily lives operate in pretty much an automatic fashion producing preconditioned responses to events. During the course of our lives we seldom recognize that this process is taking place. However, occasionally after some particularly obvious maladjusted response to some event we may wonder why we performed the action. Haven't you had this type of experience? The degree of presence in ones personality of these undesirable traits can be altered. The Buddhists have effective methods for the attenuation of afflictive states of mind and enhancement of traits that are desirable. There are many methods they have developed for this purpose, in general they consist of antidotes and the ability to recognize the appearance of undesirable states of mind in real time. By having this type of awareness the individual has the opportunity to choose to enact the desire or not. Antidotes rely upon the fact that opposites can not exist together at the same time. To give an example, feelings of anger can not coexist with feelings of compassion, etc. The capability to recognize these emotions as they appear and also have the ability to choose whether to enact them or not, is developed through meditation.

Before we can proceed further in our discussion the afflictive psychological traits and their antidotes need to be identified.[117] The psychological traits that are essential and/or conducive to supporting the operation of greed within our society are; an exalted valuation of the needs of

---

[117]   *Before starting with this list it should be noted that the context and amplitude of the expression of some of these traits govern whether they are beneficial or not.*

the individual, excessive competitiveness, an adversarial view point, a narrow simplified perception of others as ciphers that are there to be exploited or defeated if resistance is offered. Once the competitor or recalcitrant individuals frustrate the actions of the greedy self important person it often produces anger. The attributes of anger are an exclusive focus limited to a few aspects of a situation. The result of the limited focus is that the few features that are focused on become very large in psychic landscape, thereby producing distorted responses to the ignored broader situation. The antidotes for these mental states are; compassion which precludes anger, adversarial proclivities and the view of others as ciphers. Generosity eliminates greed, cooperation is antagonistic to competition, patience eliminates frustration and empathy in the context of group solidarity negates a callous outlook and exaggerated level of individualism.

The grass roots social organization that is being proposed would be comprised of a more communal style of living where the members do not privately own the facilities were they reside. These types of communities are referred to as intentional communities. In these communities members receive some personal cash periodically for personal usage and everything else is provided by the community, i.e., food, housing, transportation, recreational facilities, day care, educational support, etc. The fruits of their labor are of equal value and belong to their organization. The type of organization being proposed is a non-for-profit corporation which is formed by being Incorporated under IRS code 501 D. In this type of corporation taxable income is paid individually by the members who are considered to be partners.

The structure of the proposed community has a highly egalitarian character and is managed using the planner/manager system. In this system three planners are elected for a 1 ½ year period which is staggered. They are not allowed to continuously occupy this position for longer than one period. Their purpose is to make strategic decisions for the community based upon members (communards) input. The method the planners use is to post a proposal in a designated community area where members will consider the proposal. Upon consideration of pending proposals the members affix written and signed comments. The planners use these comments to refine the proposal. Once a proposal is in its final form it is submitted to the community for approval which requires 80 % of the community to be in favor of it for implementation. If the 80 % approval is not reached the project is shelved or dropped. The mangers take care of the operational aspects of the various activities. In these communities the managers are usually workers in other areas. The planners also participate as managers and workers; the social structure is not hierarchical. For example, a manager of the vegetable garden may be a worker in building maintenance, etc. It should be emphasized that these people occupy these positions through strong community support and can be easily removed from their positions by the members.[lxii] In the intentional community Twin Oaks which is being used as a model in this discussion a member can be removed if 20 % of the other members vote for them to leave. If you want to probe how these communities operate in depth, I would highly recommend that you read "Is It Utopia Yet?" which is about the Twin Oaks community. The Intentional Community organization and Twin Oaks community web sites also have all types of useful detailed information such as incorporation documents, membership agreements, etc.

The basic unit of the proposed new grass roots level of social organization is largely based upon this style of egalitarian community. Henceforth, I will refer to these basic units as community segments or more simply as segments. The number of persons who are members of a segment needs to be restricted to a maximum of 200 and in the early formative period to 100 or less (Twin Oaks was started in 1967 and usually has around 85 adults and 15 children according to Kinkade). The reason for the numerical restriction of 200 is based upon anthropological research on egalitarian societies. This research found that the maximum number of people that a human being can actually have any significant knowledge of and interaction with, is 200. What occurs when a egalitarian community grows any larger, the level of direct personal knowledge of the other members becomes too tenuous. Because of the excessively attenuated level of knowledge after 200 members are reached, this lack provides the latitude for persons of poor character and an opportunistic bent to engage in nefarious, undesirable, and destructive behavior to the community. For example, the Hutterites who have lived in collective communities for hundreds of years limit the size of their communities to around 100 persons. They have found that when their communities are larger someone will engage in dishonest activities. Once one person starts to engage in undesirable actions their activities then corrupt other members and the community will eventually collapse. The Buddhists have larger monastic communities than 200 persons, however the monks or nuns live a spartan life in a highly structured communal environment. Traditionally a monk or nun only posses 3 robes and a bowl and does not have any money. They also have an extensive network of training monasteries that are structured to provide an environment suited to the attenuation of afflictive states of mind. Initially the Buddhist trainee enters the monastery with all their raw abrasive undesirable egotistical characteristics. The monastic regimen is designed to chafe against undesirable states of mind causing them to manifest, thereby allowing them to be identified and eliminated. The process of attenuation and elimination is guided by senior monks who are very skilled in this form of training. A significant part of this process employs various types of meditation that allows the trainee to metabolize the evoked afflictive states of mind and augment desirable states. If you wish to delve into training and life at Buddhist monasteries you may wish to refer to D. T. Suzuki's book. While I have no personal experience of being in a monks training monastery, I have been on a number of Zen sesshin[118] (retreats) that give a slight taste of the more thorough training at a monastery. As a result of my early sesshin experiences I wrote the following descriptive poem:

Pebbles in the stream with sharp edges,

growing smooth cry; ouch, ouch!!!

Of course the segment style of egalitarian community we are discussing isn't being envisioned as being a knock-off of a Buddhist monastery. However there are some commonalities, a shared cooperative life that would tend to evoke undesirable states of mind in egotistical people. Undesirable states of mind are a source of many of the problematic actions which generate acrimony and internal pressures that disrupt the harmony of the community. An effective means of reducing the presence of negative personality traits is by the application of well proven Buddhist meditation practices. In order to effectively carry out these practices a qualified meditation teacher will be

---

[118]   *Sesshin can be translated as meaning touching the mind.*

needed to provide instruction and also provide ongoing support to members who are willing to work on personal improvement. What will happen after awhile, as these negative mental states are reduced the sharp edges will be gone and things will become smooth, smooth and easy, easy without as much ouch, ouch.

The reason "segment" was chosen to describe an individual egalitarian community is that the envisioned social structure would be comprised of a larger network of loosely interactive segments. By having this arrangement the advantages of economies of scale and a broad range of skill sets would be available. Thus, a social organization with the advantages of large scale while preserving the social advantages of the small individual segments can be attained. The combination of these characteristics have the potential to be a practical means of providing all the goods and services presently produced by the multinational laissez faire capitalist system without its environmental and social disadvantages.

If the basic economic activities that are organic to these types of communities are considered, this style of organization would produce a significantly reduced level of consumption and impact on the environment. The production of much of their own food on site eliminates squandering resources on packaging, transportation, and storage for fresh and preserved foods. Since the food needs of the low number of people in these communities are relatively small it is impractical for them to utilize large scale industrial agricultural methods. This in fact is much more productive, according to the 1992 United States agricultural census small farms grow 2 to 10 times as much food per acre as large farms and tiny farms that are 4 acres or less can be 100 times more productive than the industrial giants. A 1989 national research study also found that "well managed alternative farming systems nearly always use less synthetic chemical pesticides, fertilizers and antibiotics per unit of production than conventional farms." The same study found that the reduced use of these inputs also reduced production costs and the potential of environmental and negative health effects while not decreasing and often increasing per acre crop yields. Another 24 year study conducted in China on the effectiveness of chemical methods found that soil fertility declined unless supplemented with straw and manure. How does the GMO crops compare to small farms? Contrary to the PR about how GMO crops are needed to save the world from starvation recent studies by the National Academy of Sciences found that GMO seeds actually produced smaller harvests than hybrid seeds and a USDA study also found no overall reduction of pesticide usage.[119] A more recent 30 year study conducted by the Rodale Institute[lxiii] found that the yield of organic farming methods matched conventional and GMO type crops during years with adequate rainfall. However, the use of organic methods out performed the conventional and GMO crops during years of moderate drought.[120] The differential they found is fairly large, to use organically grown corn as an example, they found that during periods of low precipitation the average yields were 134 bushels/acre for organic, and 102 bushels/acre from conventional chemical based farming. This study also indicated that traditional plant breeding methods have increased the "yields of major

---

[119] *Pesticide usage since the introduction of GMO crops has gone up by over 370,000,000 pounds per year in this time period.*

[120] *Only the last 14 years of this study included GMO crops since they did not exist when the study started.*

grain crops three to four times more than GM varieties (p.12)." They also found that the use of GMO crops have produced a great increase in the use of herbicides and number of resistant weeds. Recently the EPA approved an increase of twenty times the allowable amount of the herbicide Glyphosate residue that can be left on food as a result of its increasing use (Rodale p.12). Obviously changing to a society comprised of the smaller types of farms used by the envisioned segmented society would greatly reduce the amount of land required to provide food. The surplus land that would no longer be needed could be returned to a natural state. By restoring the surplus land to a natural state large reductions in greenhouse gas emissions from agricultural sources would be produced, in addition to reductions in fossil fuel usage and improvements in ecological balance.

Another great advantage these types of communities have for agriculture is that the property is owned by the community corporation which provides a long term type of land tenure. By having long term land tenure it is in the segments interest to use the best agricultural practices that avoids erosion and improves soil fertility. It also provides the opportunity for the communities to practice micro-adaptation of their crops to the specific agricultural environment on their property. Traditionally plants were selected by removing the poorer ones before they produced pollen. This practice eliminates their contribution to the gene pool. The second part of this technique was to retain seeds from the very best plants for future plantings. This method produced perfectly adapted plants to the local conditions giving very high yield and superior innate resistance to plant diseases and pests. Saving seed in this way also provides the further advantage of eliminating the costs of purchasing seeds, transporting them, and the use of excessive disposable packaging.

For non-industrial livestock based forms of agriculture the segmented network system provides an ideal operational environment. Indeed, if we consider the Salatin method it requires long term land tenure and a highly adaptable ongoing developmental process based upon local knowledge. For example, at Polyface they implemented a method called line breeding (used for animals). Line breeding is a system used to amplify desired traits and reduce undesirable ones, thereby increasing animal hardiness. Let's take a closer look at how this technique works. In the area where Polyface farm is located an insect engendered cow malady called "black heel" is prevalent. It occurs as a result of the insects laying their eggs on the heels of cattle. These eventually hatch producing larva. The larva bore into the cows body near where it hatches and continues on eating its way through the animals body. The adult insects eventually emerge from the animals back and seek to find and infect new hosts. The Salatins eliminated this affliction by culling out the animals with the greatest level of infestation and by having the animals with the lowest levels of infestation to replicate. This procedure was practiced for generation after generation of cows. Over time, each succeeding generation of cows developed progressively greater pest resistance. Ultimately it resulted in cattle that are resistant to this affliction. It required thirty years to complete.[121]

This type of social organization provides a number of other significant advantages over what currently prevails. Time and resources spent commuting to work are greatly reduced since most of the communards perform their work on site. The reduction in commuting and freight transport moves us much closer to our goals of reducing fossil fuel usage.

---

[121]   *As you can see this is the same method that is used to produce highly adapted land races of plants.*

217

These reductions would be realized not only from the reduction in traffic, but also by lowering the amount of wear and tear on publicly owned roads. Road maintenance and construction uses large quantities of materials that takes a lot of energy to produce and apply. By reducing road maintenance and construction the need to collect taxes to fund this activity would also be reduced.

Since these communities engage in a broad spectrum of economic and social activities they generally have facilities that are far superior to those than is generally available to citizens of the current wider society. For example, generally these communities have a large efficient well outfitted commercial style kitchen and dining area where most of the meals are taken and a few smaller kitchens for communards who want to individually prepare a meal or snack. In these types of communities a work quota employing labor credits where one labor credit corresponds to a work hour[122] for everyone is established. So if one chooses to be a cook, wash dishes, do laundry, garden, provide maintenance, work on construction, receive medical care, computer work, etc., all of these and more produce labor credits. In these communities communards often work at more than one job in a day and can change jobs according to their desire, thereby providing much greater variety and also opportunities for personal growth. Another great advantage is that the things that need to be purchased from outside the community can be obtained at much lower bulk prices.

In general these communities do not have the capability to produce all the necessities and must participate in the wider

---

[122] *With the exception of child care where only a partial work credit / hour child care is received.*

economic system. Most of the commercial activities are comprised of a variety of small scale enterprises. These enterprises usually sell something they produce or offer some type of service. Communal manufacturing usually consists of handicrafts and cottage industries. Some of the communities operate construction businesses or other related types of services. They occasionally own and operate small retail stores, restaurants, etc. These enterprises are usually sited on their own property or in a nearby location. The types of services they offer are dependent upon the capabilities and desires of their members such as; education, medical, counseling, design, etc. Twin Oaks for example, makes hammocks, tofu, and offers cataloging services to generate cash. In some of these communities some of the communards may work outside the community to produce income.

The most fundamental capability for this type of social organization is that it must provide a social system that supports an environment that produces enhancement of desirable personal and collective traits while diminishing destructive ones. By having this type of society its activities will automatically act as a constructive force for enhancement of the surrounding system of natural processes and personal happiness. A second important characteristic of the new society is a capability to be highly adaptable in order to cope with the anticipated more variable environment. These requirements can be realized by having the individual segments participate in a loosely organized interdependent system of exchange. The system of exchange should include; material things, specialized capabilities of communards, social innovations, specialized production capabilities and new technical innovations. The networked segments can provide all this in addition to the formation of structures capable of undertaking large scale collective endeavors. A few examples might be the production of shoes, apparel , tools, or more complex types of equipment. Efficiencies of scale would originate from the large numbers of people that would comprise the network of segments. Thus, these networks would be able to produce the necessary volume requirements for efficient production in addition to the expertise and other resources to produce such items. This could be accomplished by direct exchange of labor credits or material objects between the participants or possibly in a co-op type of system. For some kinds of endeavors requiring a conglomeration of specialized skills and resources a temporary community comprised of members with the needed skills could be formed. This type of temporary community could draw on their source communities resources to accomplish a particular task and then dissolved. Another less specialized use of conglomeration could be applied to seasonal tasks where a single segment has a periodic inadequacy of personnel. At the present time some of the types of exchange just discussed is taking place between communities through the transfer of labor credits. The Federation of Egalitarian Communities (FEC) has a program called LEX (Labor Exchange) for this purpose.

In order to produce new scion segments all of these forms of exchange could come into play. By using this system the scion segment could be provided with some type of business from the parent segments. By providing this type of economic support the scion would have a solid capability to pay the parent segments back. This process of formation would also integrate the new scion in the parent network and the wider economy through the means of exchange. For things that require very large scale such as high level educational institutions, large machinery, etc., this principle could be expanded where a number of networks may pool their resources.

This type of networked system has the great advantage of not having profits extracted by the original producer and at every subsequent step of the exchange process, thereby eliminating the constant economic drain caused by the diversion of resources into the coffers of economic drones. The net effect would be to greatly reduce the costs of goods and services. The other great advantage is that all the participants in this type of system would receive fair treatment as a result of the elimination of the deceptive, extractive, excessively wasteful, and aggressive tactics that are the hallmarks of the laissez fair capitalist system. It also has the advantage of a local/regional means of the production of necessities. Having its own organic production capability also provides the additional benefit of isolating the participants from world economic and social perturbations. As discussed above these types of destabilizing forces are inherent in the global economic system and will become magnified from increasing political and environmental instabilities. A final advantage is that it is not based upon a high degree of financing. By having low levels of financing these communities sensitivity to the vagaries and demands of the prevailing financial system would be reduced.

In the United States and abroad a large number of intentional communities exist with a very broad spectrum of underlying visions for their purpose. Some of these existing communities could also participate in a network either as a regular presence or on an ad hoc basis if their vision and economic system is compatible. For example, there are a number of communities that are dedicated to providing specialized services such as; conflict resolution training, medical or wellness services, etc.

If you desire to pursue the formation of an intentional community a lot of very good work has been done on the "nuts and bolts" of the process. The organizations listed below can also provide various kinds of help from sharing expertise to attracting potential members. The FEC has listings on their web site for these purposes.

~~~~~~~~~~

Chapter 6:
Political

Let's consider how the American political system actually works. The observations that are being presented are based upon my own personal experience. I have worked in about 8 – 10 municipal elections and also on a national election for a third party (the NLP). Have you ever noticed that it's quite uncommon for an independent or third party candidate to succeed in winning an election. The reason is that it is not a level playing field for independents or new parties; the two major parties enjoy great advantages.[123] In America politics runs on money. The largest chunks of money the two major parties receive come from the owners and/or operators of big business. These businesses also provide further support by providing both material and PR for election campaigns. This is not done as a result of altruistic motives but because they expect and receive a return on their investment (remember the Ford / Dodge court case).

The two major parties have several other techniques for reducing or eliminating outside competition. Election Laws in the United States are the prerogative of each individual state. The details of these regulations vary since they are produced by the individual state legislatures. In the area where I was working an independent candidate or a new or small third party were required to collect 20 times the number of signatures to gain ballot access compared to the major parties. The rational for this was that the large

---

[123]   *I am referring to elections above the local level such as school boards etc.*

parties have primaries. When I was working on the third party's national campaign one of the states required that political parties hold a primary. In order to be able to hold a primary in this state, a party had to have gained 5 % of the vote in the last election. This requirement was obviously designed to make it difficult or impossible to gain ballot access since any new party had not participated in the last election. Thus, ballot access was denied. This brings us to the next group of strategies employed by the major parties that is making it difficult and costly to gain ballot access. Of course, it is illegal to have the type of election law for primaries described above. The NLP party challenged this in court and won which was no surprise to anyone. However, the law operated as intended,that is to use up resources. The legal expenses for these court challenges are large, thereby using up funding that would be directly applied to the political campaign. These tactics also use up party members time and delays the parties campaign. Another routine technique is to challenge the validity of the ballot access petitions. The petition format is closely scrutinized for flaws and challenged if any are found. To avoid this we closely modeled the petitions we used on the ones used by the major parties. Once the signatures have been obtained the petitions have to be signed and notarized by the signature collector, party officials and a notary public. In one of the elections I worked on some of the petitions that were collected were successfully challenged and rejected. This occurred because there were several notary publics sitting at the same table validating the petition pages and one person inadvertently picked up the other persons notary stamp and stamped the petition, thus the notary stamp did not correspond to the correct notary signature. The other common petition challenge technique is to scrutinize the signatures which can also be challenged. The signatures must written in ink, be the persons official signature (no initials), have a correctly written out corresponding address with no abbreviations and be dated correctly. To file a petition signature challenge a signature simply has to look suspicious. The objectives of this process is to eliminate enough signatures so that ballot access is denied for an inadequate number of signatures, use up resources, and cause delays as described above. In some cases even if the challenges are overcome it may be too late to have the candidate's name appear on the ballot. If this occurs then the candidate has to run as a write in, which greatly reduces their chances of success. In the elections I participated in we collected ¼ to 1/3 more signatures than required. The minimum number of signatures that should be collected is 20% above the legal requirement to survive a petition challenge of this type.

Another common means the two major parties employ is to divide and conquer. The method is to create some additional competition to split their rivals potential votes. This method was used by John F. Kennedy who had no political experience, yet he was able to defeat an incumbent senator. The method that was used was to find a person with the identical name to the incumbent senator who would agree to appear on the ballot. Petitions were circulated and ballot access was gained for him. When the voter entered the voting booth they were confronted with a ballot with Kennedy's name and two other names that were identical, the incumbent and Kennedy's man. This split the vote and allowed Kennedy to win the election. I worked on a series of municipal elections where the party I worked for was beaten soundly in the first election and in the subsequent election lost by about 150 votes, a photo finish. The third election seemed promising but the incumbent party utilized the divide and conquer strategy. Some ambitious people were found to start an additional party to divide the vote. Members of the incumbent party provided funds and help for the additional party to operate their election campaign. The vote was divided and the incumbents won the election. This maneuver relies upon the fact that the major parties have a reliable party constituency. Where this election took place their reliable voter base was around 20% out of about 45% of the people who usually turned out to vote. In the following election cycle a few of the people from the diversionary party which no longer existed were on the incumbents party ticket and succeed in winning political positions.

Next a way to reduce the state of affairs just described will be presented. First though, one should realize that as long as our societies central driving force is greed all that can be hoped to be accomplished is a reduction in the influence of money in our political system. The proposed system would utilize a non-for-profit organization that would maintain a website with capabilities designed to level the playing field for candidates. The website would provide the following services for political candidates; a venue where they could list only their positions on the current issues and a brief biographical sketch. No criticism of anything about other candidates would be allowed. A nominating petition service; the petition service would operate by asking persons using the site if they would be willing to sign a nomination petition. If this was their desire, they could leave contact information such as their e-mail address. The next step in the process would be to notify them about times and locations where nominating petitions would be available for them to sign or send someone to collect their signature. This could be done efficiently by having all the various candidates petitions in one location allowing voters to sign a variety of petitions of their choice both for political candidates and referendums. A second feature of the web site would be to provide for the voter a simplified method for candidate selection. This could be done by having a list of the current issues with a spectrum of numerical scores (possibly 1-10) representing the strength of the voters preferences for each campaign issue. The computer could compare the voters preferences to the candidates positions and produce the best overall match. This would enable the voter to separate the candidates with agreeable goals from the ones that are undesirable and simplify candidate choice for the voter. Initially the web site could be funded by a small setup fee paid by the political candidate to be listed and also by small donations from voters using the service. Eventually public financing to operate the site could be developed eliminating the need for fees.

This type of web site would also be useful to provide information to the political candidates in shaping their platforms by providing information on the degree of public support for each of their proposals. A further application of this system would be to provide a forum for the public to propose new ideas and discuss their merits and eventually develop consensus. If sufficient consensus is present the new proposal could be dispatched to the candidates who may choose to incorporate it in their platform.

A further way of reducing the influence of money and help level the playing field in the American political system would be to limit the length of time political campaigns could be conducted. For example, if the national presidential and vice presidential campaign times could be limited to six months, the candidates would still have plenty of time to visit all of the states multiple times. Senatorial political campaigns could be limited to 3 months which would be plenty of time since only one state needs to be covered. A period of 1 ½ months would be adequate for representatives. By having specified shorter periods, the costs for political campaigns would be significantly reduced. Another measure that could be taken to improve the political system would be to eliminate private funding of political campaigns. By switching to public funding each candidate would be allotted the same amount of funding, thereby producing a more level playing field. With this system there would be no need for political candidates to pander for campaign funding in exchange for later favorable treatment of large donors. Not only would this further level the playing field but also allow incumbents to spend more of their time working on legislation and managing the bureaucracy, what they were elected to do, instead of spending time fund raising for future elections. Moreover, by having publicly funded campaigns the politicians would not have the constraints of short planning time horizons that are now being used to satisfy the desire of their financial supporters whose goals are to realize quick profits.

The way voting takes place in the U.S. Congress and Senate is by 3 methods, voice votes, and roll call. Voice votes is a show of hands for yea or nay and is made by estimating the proportions, this type of voting is generally used to reinstate ongoing programs and for legislation where a large obvious difference exists for the level of support for the matter at hand. Roll call, where each senator stands up and says yea or nay which is simply tallied and this style of voting is seldom used. Teller votes where the way a particular member of congress or the senate votes is recorded. This is accomplished electronically by sliding a plastic identity card into one of three slots incorporated in the back of their seats that indicate yea, nay, or abstain. By switching to exclusive use of teller votes it would be possible to identify how an office holder actually voted, and how much of the legislation was actually voted on. This would provide the electorate a definitive means of determining what an elected official actually did. A further improvement would be a reduction of the number and size of omnibus bills. These types bills are undesirable because they are a vehicle used to incorporate a lot of pork for log rolling. Implementation of these measures would greatly increase the transparency of a politicians activities.

All of these suggestions would go a long way towards improving how the American political system performs. Unfortunately it will not completely eliminate the role money plays in our political process. There would still be lobbying[124] and the revolving door where government officials alternate between high paying positions in big businesses and influential positions in government. Even with these improvements the powerful well funded interests will continue to have a great deal of influence on the political process. Their capacity to provide their well developed PR and other capabilities to directly and indirectly shape public and office holders perspectives to their advantage (remember UFCO) will still be present.

Let's consider what effects a switch to a networked segmented society would have on the short comings that are present in our current political system. In the early developmental stages when individual segments and small incipient networks exist, the amount of political influence could be effective at the local level and through lobbying. Local political influence would develop as a result of the formation of clusters of segments. Clusters are small groups of segments living in close proximity to each other would also have the advantages of allowing the presence of

---

[124]   *Lobbying does produce useful input to elected officials for shaping policy, however, greater transparency would improve the activities of politicians. The proposed website could be a useful tool to provide this type of transparency.*

organic professional services, educational facilities for children, and the economic advantages already described.

If segmented networks becomes the predominant type of social organization then the elected officials would also be members of egalitarian communities. By being a member of a segment a politician would not have the means for accruing direct personal gain since what ever material or wealth they would acquire would just go into their communities fund. Thus, the mechanism of producing biased legislation would be virtually eliminated. Of course a politician would be able to accrue some gain by enhancing the segment he/she lives in. It is extremely doubtful that this would work though, since these types of activities would be impossible to conceal from the other members of the segment. Also each segment would be a participant in a number of interlocking networks where information, personnel and regular exchange would be taking place. By having this type of networked exchange inter-segment transparency is created. With this type of transparency corruption would be impossible to conceal from the other segments of the network for any significant length of time. The network(s) that an individual segment belongs to are also in possession of a significant lever to insure upright behavior. This source of discipline would occur automatically since an individual segment is dependent upon the rest of the network to supply things that can not be produced on site. The other segments could simply exclude the malefactor from participation in the network(s). This type of shunning would produce progressively greater hardship upon the isolated segment to abandon unsound practices. Of course the prevailing state and federal legal system would also be available to offer remedies for undesirable activities.

My assessment of the current situation is that there is much to be optimistic about. The awareness and desire to address the problems discussed in this book is growing rapidly in American society and throughout the world. It is my hope that some or all of the ideas presented may be implemented in the form presented in this book or improved upon by others. At a minimum this book should at least provide a basis for discussion about shaping a new and hopefully better society, while remediating the serious environmental challenges existing throughout the world. When viewed from a wider perspective it is obvious that the current economic system and the culture that supports it are obsolete and dysfunctional. All that is required to make the transition already exists. It is as simple as making up our mind to change and then carrying out the program. One should not look upon this with trepidation or fear, after all we are in charge and can make this opportunity bear splendid fruit and have fun doing it!

I will close this section with the poem - I'm Nobody! Who are You? written by Emily Dickinson:

I'm Nobody! Who are you?
Are you – Nobody – too?
Then there's a pair of us!
Don't tell! They'd advertise – you know!
How dreary – to be – Somebody!
How public – like a frog -
To tell one's name – the livelong June -
To an admiring Bog!

Conclusion

Attachment comes at wasteful cost:
Hoarding leads to certain loss;
Knowing what is enough avoids disgrace;
Knowing when to stop secures from peril;
Only thus can you long last.
Tao Teh Ching by Lao Tzu

If consideration is given to the technologies and agricultural techniques presented in this book it is obvious that it is entirely feasible to change to a sustainable pollution free society. The technologies and agricultural methods are available for use right now, adequate for the task, and can be accomplished at a realistic price. In fact the numbers presented herein exceeded a 100% annual reduction in greenhouse gas output. It is important to attain the 100% greenhouse reduction level as quickly as possible to stabilize climatic forcing. However, the impression that I gained while researching climatological information for this book is that a desirable level of atmospheric carbon dioxide is in the range of 315 – 325 parts per million. This means that we need to actually go beyond a 100% greenhouse gas reduction and start removing these substances from the atmosphere. This will probably take a significant length of time since vast quantities of these gasses are being stored in the oceans. This too should also be achievable if we change to a social system that is based upon conservation of resources instead of consumption. If we can manage to transition to this type of society a large proportion of the current resource usage and its accompanying pollution could easily be eliminated. At the present time 20% of the energy used and huge quantities of materials are consumed in the US by the manufacturing sector. Moving to a conservation type of economic system would greatly diminish the consumption from manufacturing as well as the use of transportation, producing further direct and indirect reductions.

The major sources of pollution in the oceans come from the air and agricultural sources from runoff. By eliminating the use of coal and switching to the agricultural methods suggested earlier in this book most of the sources of oceanic pollution would be eliminated greatly enhancing the potential of marine recovery. Termination of these two sources of pollution also has the huge benefit of stopping most of the production of methane which is a strong greenhouse gas and only remains in the atmosphere for 12 years. This would act as a fast check on global warming while the slower acting measures take effect.

The segmented network type of social system has a number of additional advantages over the current system. It is a non-hierarchical type of system that allows the rapid transmission of information directly to the appropriate individuals. These individuals could formulate appropriate responses and quickly marshal resources to deal with problems. The current system we have is comprised of bureaucracies that are geared to deal with routine predictable types of problems. Their capability to quickly respond to novel situations is poor. For example, the small scale near shore fishermen recognized that the size of the fish they were catching and the amount landed was diminishing. They correctly surmised that this was an indication that there were problems developing within the fishery. They tried to gain the attention of the authorities that there were problems in the fishery. The government ignored these warnings and did not take any action for a considerable time allowing the situation to continue to deteriorate.

Perhaps we can take a lesson from this about how to improve the management of the fisheries. One of the most significant contributors to the collapse of the fisheries is that we continue to allow too much fishing effort. Much of this overcapacity is incorporated in the large industrial ocean going fishing ships that can go anywhere in the world, clean out a fishery, and then move on to a new one. If the use of these types of vessels were abandoned and only small shorter range boats were used it would eliminate the destructive hit and run tactics that these large factory ships employ. Another approach would be to grant long term exclusive fishing rights for a particular area to a local segmented network or co-op (no foreign vessels allowed to fish). All fishing vessels could be required to have an electronic transponder to make it easy to track them to prevent encroaching and poaching. This would be an improved management approach because it would be in a network's interest to maintain the area they control in good sustainable condition.

What are a few simple things that we could do to improve the maritime environmental situation? If you are going to eat fish consider choosing farmed fresh water vegetarian fish such as tilapia. If you wish to eat ocean fish try to choose species that are not depleted and fished in a sustainable way. Information on which kinds of fish that meet these criteria can be obtained from the Marine Stewardship Council which is a certification organization for sustainable practices. The Monterrey Bay Aquarium provides information on the status of individual types of fish and their websites are listed below. Support regulatory measures that reduce fishing effort, expand marine sanctuaries and that reduce destructive practices to marine habitat. If you can, add your vegetable and fruit kitchen waste to a compost pile.[125] Most of this type of "waste" now goes into landfills.[lxiv] By putting these materials in landfills large quantities of valuable soil nutrients are removed from the soil nutrient cycle.[126] Moreover, by burying kitchen scraps in landfills they also produce methane which contributes to global warming. Join a group that is working on some aspect of this that you are interested in, or create your own organization; organize, organize, organize, network, network, network. The important thing is to invest some time in moving things ahead. All we really need to do is take that first step, and then another; the path will open before us. It is like building the Golden Gate Bridge,it was done by each person one rivet at a time. That bridge would not exist if the work done by even one person was missing.

If you are interested in finding actual information produced by climate scientists, the Skeptical Science web site is a good place to look. A link to their web site is provided below.

---

[125] According to Katz a few cities have implemented composting programs that utilize kitchen waste. He reported that San Francisco was able to reduce its municipal waste by 19%. After it is composted it is sold under the rubric "Four Course Compost".

[126] In the Humanure Handbook, Joseph Jenkins goes even further and advocates composting human feces. The thermophilic method he advocates produces compost which is completely free of pathogens. It also has the further advantages of being a fine soil amendment, and uses little water.

~~~~~~~~~~

http://www.seafoodwatch.com/ (Monterey Aquarium site.)

http://www.msc.com (Marine Stewardship Council site.)

http://www.skepticalscience.com/ (Scientific Climate Information.)

http://www.rodaleinstitute.org/ (Independent science based information on sustainable agriculture)

~~~~~~~~~~

Planet earth is very clearly showing us that it is moving from a benign state into a new mode that will be more challenging for human beings to live in. We should take heed of what is being presented to us. It should be kept in mind that planet earth can adjust to our activities and continue on without human beings. Human beings, however, need a planetary environment that can not be significantly different than what currently prevails to survive. Looked at from this perspective individually and collectively we need to discontinue and avoid engaging in environmentally disruptive practices. PR, gloss, and trying to lawyer our way around these problems will not work. **We need to look at these problems as clearly as possible. We also need to be very pragmatic and use what we have to solve these problems.** Ideologies and theories are traps that prevent us from taking effective action; they need to be abandoned or ignored. We need to get the gold dust out of our eyes it's not possible to see these problems and solve them when we are blinded like this. Some further observations; the past is only useful to draw lessons from. It should not be held onto or tried to be recreated. The past is gone and can not be changed. **<u>The present is the only time when anything can be effected.</u>** The future is created by what we do in the present. So waiting or hoping for some type of imagined future to appear by itself is a complete waste of time and another trap. Therefore, we must **act now; there is no other time.** Lets restore our planet to a beautiful healthy state that can provide all our needs indefinitely, now.

The last question is what kind of personal traits should the segmented networked society develop and support? In my view it should provide the means to produce individual and collective happiness. Therefore, I will conclude this book with what Shakyamuni Buddha said is the sources of happiness.

~~~~~~~~~~

Mangala Sutta
The Sutra on Happiness, Shakyamuni Buddha

The Buddha was living at Anathapindika Monastery in the Jetta grove when a deva (a radiant being) appeared and asked him a question in the form of a verse:

"Many gods and men are eager to know
what are the greatest blessings
which bring about a peaceful and happy life.
Please, Tathagata (the Buddha), will you teach us?"

(This is the Buddhas answer):
"Not to be associated with foolish ones,
to live in the company of wise people,
Honoring those who are worth honoring-
This is the greatest happiness.

"To live in a good environment,
To have planted good seeds
And to realize that you are on the right path-
This is the greatest happiness.

"To have a chance to learn and grow,
To be skillful in your profession or craft,
Practicing the precepts and loving speech-
This is the greatest happiness.

"To be able to serve and support your parents,
To cherish your own family,
To have a vocation that brings you joy-
This is the greatest happiness.

"To live honestly, generous in giving,
To offer support to relatives and friends,
Living a life of blameless conduct-
This is the greatest happiness.

"To avoid unwholesome actions,
Not caught by alcoholism or drugs,
And to be diligent in doing good things-
This is the greatest happiness.

"To be humble and polite in manner,
To be grateful and content with a simple life,

Not missing the occasion to learn the Dharma *
This is the greatest happiness.

"To persevere and be open to change,
To have regular contact with monks and nuns,
And to fully participate in Dharma discussions-
This is the greatest happiness.

"To live diligently and attentively,
To perceive the four Noble Truths,
And to realize nirvana1 **
This is the greatest happiness.

"To live in the world
With your heart undisturbed by the world,
With all sorrows ended, dwelling in peace-
This is the greatest happiness.

"For the one who accomplishes this
Is unvanquished wherever she goes,
Always he is safe and happy-
Happiness lives within oneself.[lxv]

~~~~~~~~~~

* Dharma is a complex term with many meanings, but briefly it
means the the general manifestation of reality, the general state of
affairs, things and phenomena. A second meaning is to cleave to
ethical behavior.

** Nirvana is the goal of Buddhism it requires the complete overcoming of the unwholesome roots – greed, anger, and delusion.

~~~~~~~~~~

Post Script

In order to provide some further timely information about some current developments relevant to some of the topics presented earlier I am including this post script.

At the time of this writing (September 2016) the United States and China have started discussions about engaging in a joint effort to develop a design for a molten salt thorium type nuclear reactor. While this type of technology potentially has some advantages over the liquid metal fast neutron reactors it currently does not exist. However, even if a practical molten salt reactor is developed, it does have some disadvantages. The greatest disadvantage (besides not actually existing) is that it would be necessary to start another new mining industry which of course would have negative impacts similar to those discussed earlier.

It should also be borne in mind that the United States has a huge amount of nuclear waste in temporary storage at the 2nd generation power plants. It will require about 1/4 million years for this material to decay into an inert form. This problem really needs to be addressed since on a human time scale this is basically forever. Luckily the waste problem can be corrected by deploying the 4th generation type of fast neutron reactors. Not only will this type of power generator produce waste that will effectively become inert in less than 500 years but will also reduce the cost of energy to a fraction of its current level. The only caveats about implementing this type of energy production is that it should be carefully sited to avoid areas prone to flooding and geological activity.

Recently some small tentative steps are being made toward implementing electrically powered freight transport using trucks. At the present time a demonstration system for electrically powered trucks (E-Trucks) is being used in

Sweden. This system uses overhead power lines to provide power to the trucks as described earlier. Their plan is to install this system on the heavily traveled roadways. The E-Trucks would be either a straight electrical system that incorporates a battery or a hybrid to enable them to leave the power lines. By using a network of electrically enabled roads the trucks could "hop" from one road with overhead power to another using on-board power. Another E-Truck initiative is taking place in the state of California USA. In California $23.6 million has been allocated to fund development of infrastructure and install a fleet of 43 electrically powered trucks. These trucks would be used for freight transfer in marine ports and would employ overhead power lines.

In Green county, Wisconsin the Salatin method of intensive rotational grazing is starting to be implemented. I know of one farmer who has this system operating and another who is installing the fencing for the paddocks and constructing the egg-mobiles. I have been purchasing some products from one of the farms. The prices are a little higher than what one would pay for similar organic free range products in a supermarket. The difference though between a typical free range organic poultry commodity product and the Salatin method raised animals is that they really are free range; that is they are out in a large vegetated field (I personally visit these places to verify what is going on). In my opinion the quality and flavor is superior to the commodity type of food found in most food retailers. This isn't surprising since almost all of animal based food sold in the US is produced in CAFO's. For example, 99% of poultry products are produced in these types of facilities. When you buy an "organic" poultry product from a supermarket chain store what you are almost always getting is something produced with organic certified feed in an industrial facility. If it also says "cage free" it means that the animals are not immobilized in tiny cages but are allowed to move around in the building.[127] If you are buying "free range" products generally it means that the CAFO building has a small door that opens into a small fenced in enclosure similar to a dog run.

The reason that the reality of how these animals are being raised is different from the impression one gets from the labels on industrial produced commodity organic free range or cage free foods. This discrepancy exists because the regulatory posture of the FDA to a great extent is shaped by the industry. The way the large food processing and industrial agriculture industries have accomplished this is by engaging in a costly, protracted, and robust campaign to obtain favorable legal definitions from public officials for their products. This process of linguistic and regulatory erosion has been going on right from the start of the establishment of the FDA and employs all the methods of PR and political manipulation described earlier. When I was a young man, in many cases the ingredients that these definitions relate to would have been described as being "artificial" and in some cases would have been classified as adulterated food stuffs. If you are interested in looking into this topic in greater depth the nutritionist Marion Nestle wrote an excellent book on this topic named "Food Politics".[lxvi]

---

[127]   The cage free facility I visited initially had 25,000 birds present. There were birds that showed signs of pathological behavior where their plumage had been pecked off. More recently the owners reduced the number to 19,500 birds.

~~~~~~~~~~

Thank you for spending your time reading this book.
J.F. Hagen

~~~~~~~~~~

Resources

The following is a listing of some sources of information on the formation of intentional communities and networking:

The Federation of Egalitarian Communities – http://www.thefec.org.

LEX is accessed from FEC.

The Fellowship for Intentional Community – http://www.IC.org

The International Communes Desk – http://www.communa.org

Twin Oaks Egalitarian Community – http://www.Twinoaks.org

New Economy Coalition – http://www.neweconomy.net/

~~~~~0~~~~~

Appendix A
CAFO Debunkers

No. 1

Corporate Stewardship

The question we will be considering is what degree of stewardship can be expected from large corporate agribusinesses?

Corporations exist for basically two purposes; to produce and maximize profits for their owner(s) and shield them from liabilities. These principals are firmly embedded in our legal system and act to restrict the level of altruistic activities a corporation can engage in. The following example of a law suite against Henry Ford clearly illustrates these concepts. At the time this took place he owned 68% of the Ford motor car company and wanted to upgrade it by using earnings produced from sales, thereby diminishing the amount of dividends. The Dodge brothers, who were also share holders in the company as well as a major vendor, filed a law suit against Ford for not paying the maximum amount of dividends. Ford lost this suit in 1919 and the legal decision read as follows: "But it is not within the powers of a corporation to shape and conduct a company's affairs for the merely incidental benefit of shareholders and for the primary purpose of benefiting others." This incident clearly demonstrates that incorporated businesses are legally obliged to maximize profits and that expenditures on stewardship to benefit others is of incidental concern.

Let's now consider the environmental stewardship of a large corporate agribusiness that expends a lot of resources to produce a wholesome corporate image. Duff Wilson published a series of articles in the Seattle Times[128] about a division of the Land o Lakes butter company named Cenex. Cenex was combining toxic waste with some other chemicals and selling it as fertilizer to farmers. This activity was perfectly legal since adding some extra ingredients and renaming it fertilizer legally changed it from a toxic substance to a non-toxic classification. The unsuspecting farmers who were using this "fertilizer" noticed that their crops were dying. After public exposure by the paper and an expensive law suit the farmers did receive some compensation from the company, however, their fields are now contaminated with heavy metals making them useless.

This is not an isolated case, large corporations are there to maximize profits. Some of these corporations add to their profit margins by knowingly engaging in a illegal activity. They do this because the forbidden activity may have a small cost relative to the amount of profit that can be made. Paying the fine is just considered part of the cost of doing business. I'm not suggesting that all corporations engage in this type of activity. In fact most of them work hard to operate in a legal fashion.

---

[128] Wilson Duff: Fear in The Fields – How Hazardous Waste Becomes Fertilizer, Seattle Times 1997

To make this point I shall relate the following case: One of my friends' son-in-law works for a major drywall producer. This company's manufacturing takes place in Mexico. He has had to go down to Mexico several times to fire managers in these facilities because they were producing drywall that had asbestos added to it. The central management of the company does not want asbestos in its products for obvious reasons and monitors its products. The lesson we can learn from this, is that it is difficult to effectively oversee multiple far flung industrial operations remotely.

What conclusions can we draw from all this? Large corporations will spend money on "stewardship" if they think it will be beneficial to their bottom line in some way. The most common reasons for doing this is to reduce legal exposure and to fabricate a facade (corporate image) to produce a favorable impression on the public.

Now let's take a look at some of the inherent structural factors present in large corporations that would tend to attenuate stewardship. The management of these enterprises are generally comprised of a board of directors and operating personnel. In most cases the composition of the board of directors is heavily weighted with the principal steak holders of the corporation. Since the directors want to maximize the return on their investment, they are inclined to direct the management to increase dividends during periods of good business. When business slacks off the operating management is often pressured to find ways to maintain profits to produce dividends. An obvious way to keep profits up is to reduce costs. If stewardship is consuming corporate resources it's usually one of first items to be eliminated. For example, CAFO's typically produce huge quantities of manure whose application is legally restricted to dry periods of weather. If a persistently wet pattern of weather occurs and the manure ponds are full, a significant level of motivation would be present to illegally spread manure locally instead of paying for expensive long distance trucking or reducing their inventory of livestock. Moreover, during periods of slack business, maintenance often goes wanting. This is particularly true for low priority items that do not contribute directly to profits. For a moment imagine that you are a manager of a facility operating with marginal funding and had to choose between purchasing a failed well pump or a waste pond leak, what would you do?

When considering all of this, it becomes obvious that the conditions that would give rise to a strong sustained commitment to environmental stewardship simply doesn't exist in large corporations. In fact it could be argued that the variable business climate combined with a corporations fundamental nature would make it unlikely that an effective sustained level of stewardship could be maintained over long period of time.

~~~~~0~~~~~

Community Impacts of CAFOS

CAFOS are generally large scale agricultural corporations and are not operated in the same way as the traditional family farms we are familiar with. Let's consider the basic nature of these large scale enterprises. The fundamental and overriding objective of a big business is to continuously produce ever increasing profits for the owner(s). Since the entire market for food production in the United States is fully developed there are a limited number of ways a corporation can expand its profits. The first method is based upon strategies for the expansion of market share.

Expansion of market share is accomplished by taking away from others. For example, let's consider the addition of a 6,000 head dairy operation to the system. Since the dimensions of the dairy market is essentially constant it means that a number of small dairies will be forced out of business to provide the market space for the large industrial dairy. The loss of small farms occurs because the mega-dairy has the advantage of large scale. If we do the arithmetic we find that for a 6,000 cow industrial dairy it would require 100 small 60 cow dairies to go out of business.

These large scale industrial operations also claim that they are beneficial to the community because they create jobs. Using industry employment data, the 6,000 cow dairy being discussed would employ around 60 people. The question though is there actually a net increase in employment? If we consider how family farms work, usually the husband and wife jointly operate the farm.

Using the example above, I will assume only 1/2 of the small family farms would go bankrupt since some of them may be able to shift to some other type of production. Thus, 50 of the small farms would be forced out of business and 100 persons would be deprived of their lively hood and would have to find new employment. If 60 of the unemployed displaced farmers are reemployed by the CAFO, we find that the CAFO would cause a net loss of 40 jobs. Since the actual number of farms and their associated jobs are being diminished by the CAFO system the chances for the remainder of these unemployed people of finding a job in their field of expertise would be bleak.

The second way these industrial agribusinesses use to increase profits is by transferring as much of their expenses and liabilities on to others. One of the principal areas where this occurs is to effect the transfer of liabilities on to the local community. Let's take a closer look at how the transfer works. If conditions arise that will produce unacceptable expenses or to cause profits to fall too low, the problem is usually dealt with simply by removing as much of the company's assets as possible, declare bankruptcy and then walk away. The community is left to deal with the environmental, social, and health problems that is left in the wake of the bankruptcy. The creditors and suppliers end up with a fraction of what is owed to them.

Product Demand?

The CAFO industries public relations material often points out that a high demand for their products exists. The statement about high demand is absolutely true. The industry often uses this fact to suggest that there is a need for having more CAFOs. However, high demand doesn't imply that inadequate supplies of the products are being produced and a need for expanded production exists. In fact if we look at the scale of government subsidies for agricultural products, it would seem that they are being greatly overproduced. For example at the time of this writing we find that the USDA had an inventory of 740 million pounds of cheese and 298.2 million pounds of butter that they purchased of excessive production.[129]

Obviously if a shortage of these products existed it would be quite simple for the government to cut back or eliminate its subsidies,[130] thereby allowing more product to enter the market. I'm sure that the taxpayers who are being compelled to pay billions of dollars to remove excess production would be quite happy to pay less taxes.

In view of these facts it seems that instead of establishing more CAFOs we should be shutting them down. This would reduce agricultural overproduction, unemployment, and the large quantities of pollutants they produce.

~~~~~0~~~~~

---

[129] These numbers are for April 2016, the most recently published by the USDA.

[130] According to an article in The Week.com, in 2013 10% of "farmers" received 75% of the farm subsidies. A few examples of these "farmers" are Archer Daniels Midland, the billionaire Ted Turner, Riceland Foods, Pilgrims Pride, etc.

No. 3

CAFO Diet

The CAFO industry consistently touts the fact that their livestock is fed grain. They innumerate the following benefits: that the grain producers and vendors make a lot of money from this. Of course these statements are completely true but they also present a very simplistic picture that culls out a whole herd of unpleasant facts about the industry.

If the conditions that the animals live in are considered, they are packed into unsanitary facilities like sardines in a can. Health care professionals, nutritionists, the World Health Organization (WHO), and the Center for Disease Control (CDC) have been pointing out that these facilities are perfect incubators for the formation, growth and distribution of virulent microbes.

Basically there are three CAFO characteristics that produce this health threat. The unsound practice of routinely administering prophylactic doses of antibiotics, changes that take place in the animals digestive processes produced by eating grain,[131] and the overcrowded conditions in these facilities.

---

[131] Cattle are grazers of vegetation and have digestive systems not well adapted for processing grain.

Biologists have found that the routine exposure of microbes to antibiotics causes them to adapt to these therapeutic agents. If this occurs these important medical tools are rendered ineffective for human use. Evidence for the appearance of these resistant bacteria already exists. Research conducted in Europe has demonstrated that a "strong causal link" exists for the appearance of Methicillin resistant *Staphylococcus aureus* (MRSA) and CAFO pigs (no American research of this type has been conducted).[lxvii] According to the CDC more deaths are now being produced by MRSA infections in the U.S. than are produced by HIV/AIDS.[lxviii] In Nickelsburg's article she reported that the five most likely diseases that may originate from CAFOs are; *E-Coli, MRSA, Campylobacter, Salmonella, and Mad cow.*

Cattle eating a normal diet have digestive systems that are not acidic but neutral. On the other hand humans have digestive systems that are acidic. The higher level of acidity in humans is thought to inhibit the transfer of non-acid tolerant cow microbes to humans, thereby protecting humans from infection by them. However, when cows are fed diets containing high amounts of grain their digestive systems become almost as acidic as a humans. These facts conspire to create a potential threat to human health since microbes have the ability to adapt to changes in their environment, i.e., to changes in acidity. Thereby, allowing these novel strains of microbes to much more easily migrate to human beings where they can cause illness.[lxix]

Most of the grain that is being fed at these facilities are produced by genetically modified organisms (GMO) crops. The basic process that is used to incorporate novel genetic material in a plant is as follows: the novel genetic material along with a antibiotic resistant gene and cauliflower mosaic virus is adhered to a very fine metal dust and shot into a pile of plant tissue. An antibiotic is applied to the inoculated plant tissue which kills all the non-antibiotic resistant plant material that doesn't contain the novel genes. The cauliflower virus is included because it produces plant cancer which forces the plant to produce a lot of the desired novel plant chemicals. So every GMO plant is resistant to the antibiotic used in it's creation.

Recent research has demonstrated that genetic material from GMO type plants has been transferred to *E-Coli* in the gut of animals.[lxx] Thus, we have another pathway that may lead to the emergence of novel antibiotic resistant diseases.

Considering all of this we find that multiple new pathways for the production of new and/or antibiotic resistant diseases are inherent to the CAFO system of production.

Dirt & Water

Since most of the grain crops produced in the United States are used by CAFOs these enterprises indirectly produce a huge impact on the soil and water that we are dependent upon. These impacts originate from erosion and agricultural runoff. Both of these are forms of water pollution.

According to Pimental the average amount of erosion taking place in the United States is around 1mm (1/25 inch) per year over 10 times it's replacement rate.[132] The economic impact of erosion "costs the

---

[132] It requires 500 years for nature to produce 1 inch of soil.

nation about $37.6 billion each year in productivity losses." Most of this eroded soil (60%) ends up in rivers, streams, and lakes.[lxxi] Silting up of our aquatic resources has a significant negative impact on fish and other species because of the changes produced by this process to their habitat.

The second source of water pollution is caused by runoff of agricultural chemicals and high level application of manure to agricultural land. In the last National Water Quality Inventory the EPA reported that 48% of stream pollution and 41% of lake pollution originated from agricultural sources.

In summation CAFOs high consumption of grain requires expanded grain production which engenders significant environmental problems (water pollution and erosion). The heavy reliance on grain also enables the creation of conditions that are producing drug resistant disease(s) such as MRSA and thereby, an elevated health threat to the community.

~~~~0~~~~

Manure Fires

Let's consider a few examples of these types of fires. Industrial style pork facilities are basically comprised of a large building with a floor made from slats and a manure pit underneath. The pig droppings fall upon the slatted floor and eventually make their way through the floor slots into the pit directly under the floor. These pits are usually quite deep (commonly around 30 feet). For example, the online journal Better Farming ran an article about a hog production facility about a manure gas engendered fire that killed 1,500 pigs and injured a worker. The worker was severely injured and required hospitalization for severe burns.[lxxii] In addition to the flash fires there have also been numerous instances of gas explosions. These explosions have caused varying degrees of structural damage from minor to severe as well as loss of life and injuries.

~~~~~0~~~~~

On April 22 President Obama signed the Paris treaty that commits the United States to reducing green house gas emissions. According to the EPA 9% of the US green house gas emissions are directly produced by the agricultural sector. Since agriculture is a major source of emissions it seems likely that this sector will have to produce some of the reductions.

Probing the agricultural sector's sources of emissions more deeply it becomes obvious that the CAFO agricultural system is the major generator of these gasses. The production of these gasses occurs directly and indirectly at the facility. The direct production of methane, a green house gas 25 times stronger than carbon dioxide, is produced in the waste ponds. It is also produced by the digestive processes of cattle, goats, and sheep.

Nitrous oxide and carbon dioxide are produced indirectly. Nitrous oxide is 298 times stronger and accounts for 5% of the green house gas emissions produced in the US according to the EPA. Nitrous oxide originates from ground microbes that feed on nitrates that are applied to fields in the form of chemical fertilizers for conventional crop production and by the application of manure to fields. The other major indirect source of agricultural green house gas is carbon dioxide. There are several sources of carbon dioxide from fossil fuel energy directly consumed by these facilities and that used to produce feed. A further source of carbon dioxide is a result of oxidation of exposed soil, a part of the process used in conventional methods of crop production. Since the feed used in the CAFO system is almost entirely comprised of grain it's production is a significant contributor to the air pollution they produce.

If we consider two of the major CAFO industries oil usage we find that 284 gallons of oil are consumed to produce a CAFO steer[133] and CAFO hogs require 70 gallons per head.[134] According to the USDA around 34,000,000 head of cattle and 106,900,000 hogs were slaughtered last year.[135] Using these numbers we find that 17.137 billion gallons of oil were consumed by these two CAFO industries which produced about 585 million metric tons of carbon dioxide. This is the amount of carbon dioxide produced by 88.4 million cars!

We also often hear that the methane from the ponds could be captured and used for fuel. Methane is an excellent fuel, however, when you burn methane it is converted to carbon dioxide. Thus, you are changing a powerful green house gas that stays in the air for 12 years to a weaker one that will take several centuries to fall by 3/4 with the remainder staying for thousands of years; not a good trade.[136]

The question is are there effective safe alternatives to the CAFO system? The answer is obvious. We can shift to an emphasis on modern grazing methods with supplements.

---

[133] Pimental David, et al.: The Potential for Grass Fed Livestock: Resource Constraints: (Science, Vol. 207, 1980).

[134] This value comes from a non-independently peer reviewed internet source.

[135] According to Harpers magazine CAFO's produce 99% of these meat sources.

[136] Caldieria: (Nature, Reports Climate Change, 20 November 2008), Internet On Line.

Grazing has been used for thousands of years and is environmentally benign as long as it is managed properly. By using this system all the gasses produced by waste ponds and almost all of the nitrous oxide is eliminated. It also virtually eliminates the carbon dioxide originating from soil exposure and the excessive use of fossil fuels.

The CAFO system is also dependent on large taxpayer supported subsidies to remain viable. In essence our taxes are being used to prop up an outmoded agricultural system that is highly destructive to the environment while other environmentally benign systems go wanting.

~~~~~0~~~~~

Food Tips
Or how to avoid ersatz foodish products.

1. Avoid buying things that have a lot of ingredients with names that don't sound like normal food. To test if it sounds like normal food, Michael Pollan suggests asking a third grader about it, if they cannot pronounce the name of the ingredient in question avoid it.[lxxiii]

2. If the food or its ingredients wasn't available a hundred years ago don't eat it.

3. Buy at farmers markets and ask questions about how the food was produced. Farmers will be happy to talk about their products. If they are reticent be careful.

4. Bring cash. Farmers are not large retail outlets and seldom have credit card processing equipment. Besides credit card companies charge a fee on every transaction which drives up the price.

5. Ask the food grower to see if you can drop by their operation and look things over. If they won't allow you to see what they are doing don't buy from them.

6. Bring a cooler with ice if you are going out to a farm for a food pick up. This is particularly important if you are buying fresh animal products.

7. If you go out to a facility **use your nose.** If the place stinks be thorough in your inspection. Keep in mind though that some odor at animal outfits is common (farm animals are not potty trained). Use your judgment and also your eyes.

8. To test if its ersatz place the food on the counter. If it doesn't rot in a normal length of time don't buy it again.

9. Pollan (p. xv) sums up what a good diet should be comprised of he says: "Eat food. Not too much. Mostly plants." This is excellent advice to follow. If you wish to probe into the food industry in greater depth he also wrote a book named "The Omnivores Dilemma", which is worth reading.

~~~~0~~~~

J.F.'s Awesome Diet System

Here is my super duper diet:

Eat a little of everything.

Eat at a table only.

Eat with family and/or friends if possible and avoid multitasking with TV, telephone, computer, or in a car.

Pay attention to what you are eating. You will be surprised, food actually has flavors.

Avoid drinking pop.

Stay away from food with synthetic ingredients.

Learn how to cook. If you don't know how, it's not difficult.

How I Lose Weight

I eat what I normally would for breakfast and dinner using real food. I have a **light** lunch comprised of some fruit or melon and maybe a small piece of cheese.

I get the portion size right by adjusting the amount of food for lunch so that I am hungry for an hour or two before dinner.

No snacks.

It's OK to deviate from this on **special occasions**. Special occasions are events like Thanksgiving, dinner parties, etc.

With this system I lose 1/2 – 1 kg (1-2 pounds) per week.

ⁱ Bibliography

i Elk Black: Black Elk Speaks,

ii   Hallowell, Irving. A: Ojibwa Ontology, Behavior, and world View , Teachings from the American Earth, Edited by Dennis Tedlock and Barbara Tedlock.

iii   Banyan, Thomas: The Essence of Hopi Prophecy, ausbcomp Black.com

iv   Okumura, Shohaku: Living by Vow (Wisdom Publications, 2012) P.56

v   Bodhi: The Way to End Suffering, (accesstoinsight.org) Ch.3

vi   Lindstrom, Martin: Brandwashed (Random House, 2011) P.70

vii   Public Papers of the Presidents, Dwight D. Eisenhower, 1960, p. 1035-1040

viii   Tuchman, Barbara W. The March of Folly, G.K. Hall & Co. Boston 1984

ix   McCann, American Company, P. 45

x   Cole Blasier, The Hovering Giant: U.S. Responses to Revolutionary Change in Latin America, (University of Pittsburgh Press, 1976), pp.55-56; Ronald Schneider, Communism in Guatemala 1944-1954 (New York; Prager, 1959), p48.

^{xi} Schlesinger, Stephen & Kinzer, Stephen: Bitter Fruit,(Doubleday, 1990) pp 86-87.

^{xii} McKillop, Heather: The Ancient Maya, (ABC-CLIO inc., Santa Barbara, 2004) pp. 122-129.

^{xiii} Webster, David: The Fall of the Ancient Maya, (Thames and Hudson, New York, New York , 2002) p.301.

^{xiv} Montgomery, David R.: Dirt, (University of California Press, 2007) p. 137

^{xv} Ishikawa, Eisuke: The Edo Period had an Ecological Society, (Kodansha Publishing Co., 2000)

^{xvi} Totman, Conrad: The Green Archipelago; Forestry in Preindustrial Japan, (University of California Press, 1989).

^{xvii} Marten, Gerry: How Japan Saved its Forests: The Birth of Silviculture and Community Forest Management,
pp. 1-4: excerpted from Environmental Tipping Points, (Journal of Policy Studies (Japan), No.20 (July 2005)

^{xviii} Iwamoto, Junichi: The Development of Japanese Forestry, (Internet Source).

^{xix} Diamond, Jared: Collapse, (Viking/Penguin press, 2005) p. 305.

^{xx} McClain James, L: A Modern History of Japan, W W. Norton & Co., New York, 2002) p. 131.

[xxi] Greene, D.C.,: Correspondence between William II of Holland and the Shogun of Japan A.D. 1844, (Transactions of the Asiatic Society of Japan 39 (1907), pp. 110-115.McClain James, L: A Modern History of Japan, W.W. Norton & Co., New York, 2002) p. 131.

[xxii] National Snow and Ice data Center: Contribution of the Cryosphere to Changes in Sea Level, 6 February 2010.

[xxiii] Feely, R. A., Bulletin of the American Meteorological Society, July 2008. ; Data from Mona Loa & Ocean Station Aloha.

[xxiv] Koenig, Seth: Dead Muds, (Bangor Daily News, 7 October 2011).

[xxv] EPA (2010), Methane and Nitrous Oxide Emissions from Natural Sources

[xxvi] Dressler, Andrew, et al.: American Geophysical Union (excerpted from NASA web site) 18 November 2008.

[xxvii] (2010), Methane and Nitrous Oxide Emissions from Natural Sources

[xxviii] Kusky, Timothy: Climate Change, (Infobase Publishing, New York, New York., 2009) p 49.

[xxix] Francis, Jenifer A. & Vavrus, Stephen J: Evidence linking Arctic amplification to extreme weather in mid-latitudes, (Geophysical Research Letters, Vol. 39, L06801, March 2012) p 1.

xxx   United Nations Intergovernmental Panel on Climate Change 2014.

xxxi   Canadian Natural Resources: Internet source.

xxxii   Walliser, Jessica: Good Bug Bad Bug, (Tien Wah Press, 2011) p ii.

xxxiii   Pimental, David: Journal of the Environment & Sustainability Vol. 8, 2008.

xxxiv   Peng S. et al., : Proceedings of The national Academy of Sciences: 101:9971-75, 2004.

xxxv   Davis, William: Wheat Belly, (Rodale Press, 2011).

xxxvi   Jensen, Derrick & Draffan, George: Strangely Like War, (Chelsea Green Publishing Co., 2003) p 45'

xxxvii   Earth Policy Institute: Washington D.C., 2012 (Internet Source).

xxxviii   Graham Michael: The Fish Gate, (Faber and Faber 1943).

xxxix   Rosenberg, Andrew A., W. Jeffrey Bolster, Karen E. Alexander, William B. Leavenworth, Andrew B Cooper, Matthew G. McKenzie: The History of Ocean Resources; Modeling Cod Biomass Using Historical Records. (Frontiers in Ecology And Environment 3, no.2, 2005).

xl   Pauly, Daniel & Maclean, Jay: In a Perfect Ocean, (Island Press, 2003) p 31.

[xli] Coroso, P., Kramer, M., Addiss, D., Davis, J., Haddix, A.,: Costs of Illness in the water born cryptosporidium outbreak, Milwaukee, WI., (Emerging Infectious Disease 9 (4), 2003) pp. 428-31

[xlii] Pimental David, et al: The Potential for Grass Fed Livestock: Resource Constraints, (Science, Vol. 207, 1980).

[xliii] Diamond, Jared: Collapse: How societies Choose to Fail or Succeed, (Penguin Group, 2005) pp 456-7.

[xliv] Epstein, P. R., Buonocore, J. J., Eckerle, K., Hendryx, M., Stout III, B.M., Heinberg, R., Clapp, R.W., May, B., Reinhart, N. L., Ahern, M. M., Doshi, S. K., Glustrom, L. : Full Cost Accounting for the Life Cycle of Coal, (Annals of the New York Academy of Sciences, 2011)

[xlv] Freese, Barbara: Coal a Human History, (Pegasus Publishing, 2003) p 171.

[xlvi] Oreskes, Naomi & Conway, Erik M.: Merchants of Doubt, (Bloomsbury Press, New York, 2010)

[xlvii] Archer David: The Long Thaw: (Princeton University Press, 2009) p. 11

[xlviii] Lin Tae Kwan: Geothermal Heat Pump Systems for Residential Houses; (Master of Science Thesis, University of Michigan, 2014)

xlix   Blees, Tom: Prescription for the Planet, (Booksurge.com, 2008).

l   Keith, David & Adams, Amanda: Rethinking Wind Power, (Journal of Environmental Research Letters, 2013 ; Internet Publication)

li   Beyea, J., Nicues, JW., Susser, M.: Cancer Near Three Mile Island Nuclear Plant, (American Journal of Epidemiology, 1990).

lii   Charpak, Georges, Garwin, Richard L.: Megawatts and Megatons, (University of Chicago Press, Chicago, 2002).

liii   Cravens, Gwyneth; Power to Save the World, (Alfred A. Knof, 2007) p. 211.

liv   Hannum, William H, Marsh, Gerald E, Stanford, George S: (Scientific American, December 2005) p. 87.

lv   U.S. Agriculture and Forestry Greenhouse Gas Inventory: 1990-2008. Climate Change Program Office, Office of Chief Economist, U.S. Department of Agriculture. Technical Bulletin No. 1930, June, 2011.

lvi   Lal, R.: Soil Carbon Sequestration to Mitigate Climate Change, (Internet Source) p. 13.

lvii   Salatin, Joel: The Sheer Ecstasy of being a Lunatic Farmer, (Chelsea Green Publishing, 2010)

lviii   Salatin, Joel; Pastured Poultry Profits, (Polyface Inc. 1999) p. 20

lix   Encyclopedia Britannica, online 2016.

lx   Wilson, Edward O: Half Earth, (Liveright Publishing, New York, New York, 2016) p. 186

[lxi] Ordman, Alfred: Personal Communication, 2014.

[lxii] Kinkade, Kat: Is it Utopia Yet?, Twin Oaks Publishing, 1994.

[lxiii] Farming Systems Trial, Rodale Institute, 2014.

[lxiv] Katz, Sandor E.: The revolution Will Not Be Microwaved, (Chelsea Green Publishing, White River Junction, Vermont) p. 317

[lxv] Nhat, Hanh Thich, et al: Translation from the Pali Canon: Awakening of The Heart, (Parallax Press, 2012) pp. 497 – 498

[lxvi] Nestle Marion, Food Politics: how the food industry influences Nutrition and Health, (Berkeley: University of California Press, 2002)

[lxvii] Hannah C, et al: Pigs as a Source of Methicillin-Resistant Staphylococcus aureus cc398 Infections in Denmark, (CDC, Vol. 14 No. 9, September 2008)

[lxviii] Nickelsburg Monica: 5 modern diseases grown by factory farming, (theweek.com, November 7,2013) p. 4

[lxix] Ibid, p.3

[lxx] Bauer F, Hertel C, Hammes W.P.: Transformation of Escherichia-coli in Foodstuffs, (Syst. Appl Microbiol, 1999, 22(2)) pp. 161-168.

[lxxi] Pimental David: Soil Erosion: A Food and Environmental Threat, (Journal of the Environment and Sustainability, Vol. 8, 2006)

[lxxii] Better Farming, October 1, 2012.

[lxxiii] Pollan, Michael: Food Rules, (Penguin Books, New York, New York, 2009) p. 17

Index